I0467745

THE UNIVERSE
Yet another model of space and matter

Bjørn Ursin Karlsen

ii

Preface

My aunty knew lots,
and called them crack-pots.
(1883 Broadside Ballad)

This book is born out of sheer stubbornness. I simply refuse to accept that my brain is not good enough to understand the physical world around me, and since I have never understood the theory of relativity – not to mention Quantum Electro Dynamics, which even Richard Feynman in one of his books (QED, page 9) uttered that nobody could understand – so I need to go back to pre-Einsteinian physics and take it from there.

I don't, however, challenge the established physics which I see as an impressive *top-down* approach towards a complete understanding of nature. What I am trying to do is to start from the physics of Newton, Cauchy, Navier, Maxwell, Lord Kelvin and others, not forgetting the great mathematicians of that time, and trying to merge it with later discoveries in order to get a *bottom-up* approach to basic physics.

As I see it, Nature can be compared with a scene. We can only observe what is performed on stage, but are forbidden to see what is going on backstage. Modern scientists have become very clever to set up the stage in many different ways and then to predict the outcome. Perhaps that is the only reality there is, but some of us think there are something going on backstage, and this book is an attempt to have a forbidden peep behind the scenes. It is for them this book is written.

The book consists of five parts and each part, except the first, is divided into three chapters. The first chapter in each part describes the main issues of the discussed topic in a non-mathematical way and may be read separately. The subsequent chapters are attempts to dig a bit deeper into the subject and will occasionally imply some more mathematics. The first part in the book is about some ideas on how the universe came into being. In the second part I show how properties in an elastic spatial continuum of infinite extension can be written in a form that is identical with how electrodynamic equations are written in its most general form. The third part deals with how matter can be represented as chains of oscillating nodes with an excited singularity at its core. The fourth part describes how gravity is an effect of confined disturbance energy in an expanding space, and the final part is about special and general relativity and shows how Einstein's equation can be derived from the stress energy tensors of solenoidal and irrotational deformations.

This work is wholly mine, so all the errors and crazy ideas fall entirely on my own shoulders, there is nobody else to blame. My only purpose is to put forward some – perhaps crazy – ideas of my own. Where I have found basic facts and developments, I have used them scrupulously instead of trying to prove them myself. That said, I have done my best to verify such facts by double checking them against several sources, mainly on the internet. That is because I have done my work from home without access to a proper library. After all, this is more like a presentation of an idea of mine than a scientific work. I could perhaps dig a bit deeper, but I think this has got to be enough for now, and besides I think it is due time for me to be done with it.

Finally I want to send an apology to the giants of the past who I disrespectfully have tried to crawl onto the shoulders of; and thanks to the staff behind Wikipedia and all those who put their knowledge to free use on Internet. Also thanks to a friend on the net (who wants to be anonymous) for his proposition that I should refer to space by the term 'Spatial continuum' instead of more general terms like 'Elastic medium', or even the older term 'Aether' which is burdened with a lot of different interpretations – and last but not least – to my wife for her patience when I got lost in what I used to call 'My hobby'.

About myself: Retired teacher, no scientific degree, born 1930, Hasvik, Norway.

Bjørn Ursin Karlsen

Contents

Part I

Cosmos

Chapter 1

The Universe

In the beginning when God created the heavens and the earth, the earth was a formless void and darkness covered the face of the deep, while a wind from God swept over the face of the waters. (Genesis 1:1-2)

At the end of the nineteenth century most scientists tended to believe that electromagnetic waves, and hence light, had to propagate through an aether with some elastic properties that could be described by the theory of elasticity. It was harder to get a workable conception of how matter could move without friction through such a medium. What I want to do, is to assume that space is an elastic continuum, which I will call *The spatial continuum*, and that light and matter simply are deformational waves in that continuum. In order to explain gravity, I have to assume that the spatial continuum is uniformly expanding in all directions, so somehow a Big Crunch has got to precede the Big Bang when all matter was created, and from where the spatial continuum has expanded ever since.

1.1   The Big Crunch

In this model space is an elastic continuum of infinite extension, and matter is nothing but confined disturbance energy. How energy can be confined and moreover act like point-like entities, is the topic of this book, but first sufficient amounts of energy has to be concentrated in some limited area, namely the Universe. To get any further, we have to revert to the instant when all matter was created – the Big Bang – and even to what might have preceded it – the Big Crunch. If nothing else, there was an enormous amount of energy involved in the Big Bang. Where the energy came from, is hard to tell because creating something from nothing is only for gods. But perhaps it is possible to say something about how it came together.

There is a phenomenon called *sonoluminescence* that perhaps can give us a clue. The effect was first detected in 1934 by H. Frenzel and H. Schultes who in a work on sonar, discovered that bubbles in a fluid emitted light when they turned on the ultrasound, but the discovery was not followed up until 1989 when Felipe Gaitan and Lawrence Crum created sonoluminescence from a single bubble.

By introducing a strong sound wave into a liquid containing a small air bubble, the liquid pressure will oscillate in step with the pressure in the sound wave. When pressure in this way falls, the bubble grows to a maximum, but when the pressure again builds up, the bubble undergoes a dramatic volume reduction. The radius reduces to a hundredth and the volume to a millionth of its normal value. This violent implosion continues until the pressure inside the bubble becomes 100,000 times the atmospheric pressure, and a temperature high enough to emit light, with frequencies up to the ultraviolet part of the spectrum

Let us try to figure out what might happen if an imploding shock wave containing a huge amount of energy for some reason should occur in the spatial continuum. The imploding shock wave would increase in amplitude as it approaches the focal point of the implosion. At the same time the compression would lead to an increased inertial density that slows down the wave-speed. Just like a tsunami slows down and raises to an immense wave as it approaches shallower waters near a shore, the backwards part of the compressional wave would catch up with the foremost part and utterly increase the pressure. Only our imagination can possibly tell

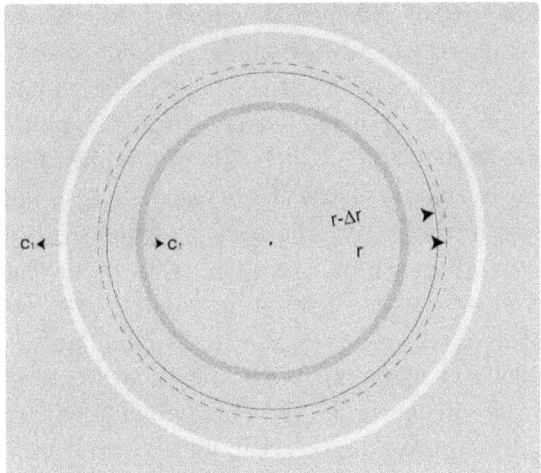

Figure 1.1: Implotion. A sudden reduction, Δr, in the radius of an imaginary sphere with radius r in the spatial continuum leads to an imploding compressional wave travelling towards the center of the sphere.

us anything about what happens when the wave eventually slams into the focal point where the mathematics just would collapse into a singularity.

Before we leave the big crunch, let us have a look far beyond it in order to get hold of what space really may be like. It is supposed to be an elastic continuum, but not an elastic stuff like steel etc. Steel consists of atoms and molecules that are bound together by stretchable bounds, and it gets its inertia from the mass of these particles. Space, however, is assumed to be a true continuum that resists deformation. But then: Where does it get its inertia from? We are all familiar with the idea that energy and matter are two sides of the same coin, and if we follow up this idea, we can try the idea that the spatial continuum gets its inertia from the energy it is coming to contain, or more precisely from the energy of being compressed. Thus space has no inertia before energy is put into it.

Mechanical waves are oscillations between potential and kinetic energy, but there can be no kinetic energy in a medium with no mass density. Hence, waves cannot travel in a space without inertia. But even the small amount of energy due to the deformational energy of the wave itself, is supposed to supply space with the necessary mass/energy density for waves to move. Wave speeds are inverse proportional to the square root of the spatial density, so the waves in the primordial space, where the deformation energy is almost nil, would move with an almost unlimited speed. If one wave should encounter another wave, however, the joint energy would increase inertia and slow down the wave speed a bit. It is a well-known fact that waves are bent towards areas with lower wave speeds, so what small energy density there initially was, the energy would tend to clump together. When an approaching wave passes through this area and is about to leave, it will pass from a dense area to a less dense area and some of the energy will be reflected back again. So wave energy tends to pulsate between outwards and inwards movements from and to an area with accumulated energy, while energy from far away is continually approaching and caught up in the area. After a long time the energy density becomes high enough to catch up all approaching energy. The area becomes like a black hole that sucks up energy from everywhere. This is a process that may go on for aeons, and if the pulsating compressional waves gradually takes the form of an almost perfect spherical shell, it might end up as the Big Crunch outlined above.

1.2   The Big Bang

The inwards and outwards moving pressure waves gradually becomes more shell-formed and in the end the compression is not reversed into an expansion, but is smashing into an almost perfect singularity – the Big Crunch. The singularity still has small irregularities such that the implosion ends up, not in one, but in a cascade of singularities in the course of a very short time interval. Each singularity would soon come to a point where the compression could go no further and the implosion would be reversed into an explosive expansion. Because spatial elements would get a high outwards directed velocity, the expansion would not stop until the volume around the singularity would be almost completely evacuated, whereupon it would again collapse into a new compressed singularity and so forth. A single entity of this kind would emit wave energy

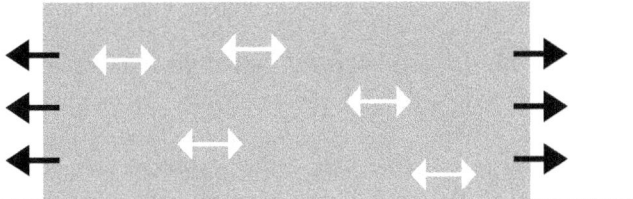

Figure 1.2: The Universe.
A schematic cross section of the density distribution of the expanding Universe. Black arrows indicate that the border of the Universe is moving outwards as a solitary wave, and white arrows that the interior of the Universe is uniformly expanding.

and hence oscillate for only a brief period, but an interchange of energy between nearby nodes might delay, or even stop, the energy loss. Each singularity would come to a point where the compression could go no further and the implosion would be reversed into an explosive expansion. If that is what happened, a small volume of highly compressed space became filled with a boiling inferno of singularities oscillating between compression and rarefaction.

However, there is also a possibility that singularities might be formed in another way, namely by inwards moving solenoidal waves. Think of a small rigid sphere that rotates back and forth around an axis through its center. Such a set-up would create an outwards directed rotational wave, but it also opens for the possibility that inwards moving waves of the same kind might occur. So nodes of this kind, or even combination of the two kinds, have got to be taken into account and added to the primordial soup of the Big Bang. Thus the plethora all this happenings became the seed from which all matter was built.

So in this model primordial matter recedes in a highly compressed space with a limited extension, which tends to expand in all directions. However, if the expansion shall not get out of control, there has got to be a condition at the border of the newly created compressed space that controls the expansion. That con-

dition could be like a *solitary wave*, which can occur in connection
with all kinds of wave movement when dispersive and non-linear
effects are balanced causing a wave pulse to propagate without dis-
tortion. Solitary waves are seen in some funnel-shaped estuaries
when the tide initializes a *bore*. A *tidal bore* is a wave that can
travel upstream in certain rivers with a comparatively slow speed,
the most famous of which is the Black (or Silver) Dragon of the
Qiantang River in China, but also in several estuaries in England
(e.g. Severn) and others around the world[1]. The bore itself has a
very limited extension in the direction of movement - almost like a
single wavelength with a marked increase in water level (up to 8.9
meter in the Black (or Silver) Dragon) - and it travels upstream
with a comparatively slow speed (typically 40 km/h). In front, the
river is flowing downwards in a natural and undisturbed manner,
but when the bore has passed, the current is reversed into a rather
smooth upstream motion, and the difference in the water levels in
the front and behind the bore is significant. So the compressed
area of space might be surrounded by a solitary wave leaving the
pressure inside the shell uniform while the pressure at the outside
remains almost nil. In this way the bore at the edge of the Universe
is moving outwards with a limited speed, so the spatial continuum
is continually expanding, and all what is inside, is following suit,
i.e. expanding (see Figure 1.2).

An interesting consequence of this pressure drop is that no waves
can ever escape from the area inside the bore. As the density at
the inside falls to almost nil at the outside, there is a very strong
density gradient between the inside and the outside, and according
to common optical principles, all waves trying to escape will be
reflected. Hence the Universe at the inside is effectively sealed off
from the outside world.

The over all picture of the Universe is that it is a compressed
part of the spatial continuum uniformly expanding in all directions.
In their book, GRAVITATION by Charles W. Misner et al, page

---

[1]Many of these spectacular phenomena are now tamed by gigantic barriers
and may not be seen any more. There are lots of pictures and descriptions of
tidal bores to be found on the internet; try to Google 'silver dragon tidal bore'.

719 [11], they raise the question: "What expands in the universe, and what does not?". They state that the expansion is between galaxies, but not on an atomic level. They compared it with the inflation of a balloon covered with pennies. The balloon would stretch and the distances between the pennies would increase, but the pennies themselves would remain unchanged. In this model, however, the expansion has to be on all levels. If it were not so, it would be impossible to explain gravity, which will become clear when that topic is being discussed.

If this model is correct, we must assume that all matter and thereby the stars and galaxies roughly will follow along with the expansion of space, but that does not imply that matter, as we now know it, is completely at rest in space. The earth moves around the sun, the sun moves in the galaxy, and The Milky Way itself moves in space. Recent studies of the cosmic background radiation have revealed that the waves are coming in with remarkably equal strength from all directions of the sky, but that is only when one has corrected for a Doppler shift of about 370 km/s relative to the radiation[2]. By taking into account the relative movement of the sun in the galaxy, one can deduce that The Milky Way as a whole moves with a speed of about 550 km/s in relation to the background radiation. If we assume that the background radiation is isotropic in relation to the spatial continuum, it would mean that The Milky Way moves with an absolute speed through space that amounts to approximately 0.2% of the speed of light. It seems like the cosmic microwave background (CMB) provides us with a fixed reference frame in sharp contrast to the relativity theory, which only opts for relative speeds between bodies. However useless it is for practical purposes, it all the same is of some philosophical interest.

---

[2]Dipole anisotropy discovered by Edward Conclin in 1969 [21] and later confirmed by the COBE Project Team

Part II

Space

Chapter 2

Light

Then God said, "Let there be light"; and there was light. (Genesis 1:3)

The key to understanding nature is first to understand light. So, what is light? A historical review tells us that almost every stone has been turned and that the only possible conclusion is that light at the same time is both particles and waves. If it is waves, then what is waving? In this chapter I'll first disregard the particle nature of light, which will be discussed later, and instead discuss how light might be waves in an elastic continuum. This leads to a comparison between the linear theory of elasticity and electrodynamics. It turns out that the two disciplines can be expressed in exactly the same form.

2.1   A short historical review

The first modern attempts to finding out what light really is like, begun in the seventeenth century with a discussion between Huygens and Newton on the subject. Huygens proposed that light is a kind of wave movement in a medium he called *ether*, but Newton got the upper hand for the next 100 years to come, with his particle theory of light. In 1705 he published a book, *Optics*, where he proposed that light came in the form of small particles, which he called *corpuscles*. They were supposed to be very small and should move with a tremendous speed. When they hit a reflecting surface, they would bounce off just like a ball hitting a wall. He explained refraction as a force of attraction speeding up the corpuscles as they entered a transparent medium, such that they were deflected towards the interior[1]. Other properties of light were difficult to explain by the particle theory. Polarization could possibly be explained by assuming that the corpuscles were not round like balls, but more like flat discusses. But other observations, like splitting up of light into different colours when passing through a prism, where harder to explain.

In 1801, almost a hundred years after Newton's *Optics*, Thomas Young performed his famous double slit experiment, also called *Young's experiment*, where he showed that monochromatic light from heated sodium passing through two narrow slits produced a very different light pattern on a second screen placed behind the double slit's screen, compared to what it did when passing through only one slit. There was no way to explain this observation with the particle theory, but Young gave a very convincing explanation by taking up Huygen's wave theory from the eighteenth century. He demonstrated both practically and theoretically how different waves, such as surface waves on water, behave in a similar way. So, for the next almost hundred years, light was considered to be waves in some medium filling up all space, even if there still were many difficult problems to explain, mainly the one that concerns how the entire earth and all other objects could pass through the ether without any resistance.

The most convincing argument for the wave theory of light came in the 1860th when James Clerk Maxwell put forward his four equations that explained all electromagnetic phenomena in one go. Almost immediately others, especially Heinrich Hertz, saw that the equations led directly to the prediction of a form of wave motion

---

[1]That part of the theory got a final blow in 1850 when Foucault was able to measure the speed of light in air and water and found that light in reality travels slower in water than in air.

which, later became known as electromagnetic waves. Electromagnetic waves came to encompass everything from radio waves through light and all the way up to gamma rays. The case seemed to be settled. Light is waves — but what is waving? So the luminiferous aether was invented, and Michelson, later accompanied by Morley, set out to detect the aether wind that necessarily had to pass by as the earth moves through space on its voyage around the sun. The result was negative; no aether wind could be detected. For the next 30 years or so, and occasionally up to this day, similar experiments have been conducted, and always with the same negative result.

Another blow to the wave theory of light came when Plank in studying black body radiation found that the pieces only fell into place when the radiation was set to come in discrete quanta, and the nail into the coffin was set when Einstein on his work on the photoelectric effects showed that light really came in packages which he called *photons*, and so we were back to the particle theory of light. The wave properties of light, however, was so well established that neither of the two could be discarded, so we ended up with the assumption that light at the same time is both particles and waves. Roughly speaking, the 18th century belonged to the particle theory of light, the 19th century to the wave theory, and the 20th century to the conception that light is both particles and waves.

That light at the same time is waves and particles seems to be almost completely inexplicable. Almost, but not quite. A wave can for example travel along a tight string fastened between two fixed points, which makes it almost point-like when seen along the direction of the string, and wavelike when seen from a direction abeam of the string. If a photon behaves like a filiform (threadlike) train of waves, it can well kick lose a single electron from a metallic plate in a photo multiplier, or leave an exposed point on a photographic plate, at the same time as it shows wave-like properties. So as a first approach, I shall stick to the assumption that light is some kind of wave motion, and leave the particle aspect to a later discussion.

2.2   Waves, but in what medium?

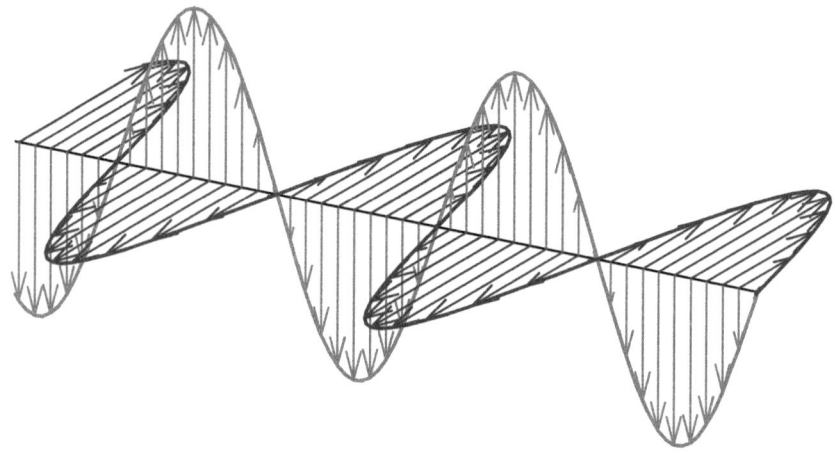

Figure 2.1: Electromagnetic wave.
A plane electromagnetic wave travelling to the right. Vertical arrows show the electric field components, and the horizontal arrows normal to the electric field, shows the magnetic field components.

Let us start with the assumption that electromagnetic waves are waves in some stuff, which we could call the *spatial continuum*. That should be general enough, and we are not burdened with old conceptions like ether, quintessence, and even space-time. The spatial continuum fills all of space because light can travel everywhere, and it should be a continuum because space seems to be the same everywhere and in all directions. Next, because electromagnetic waves, like radio waves at a great distance from the antenna, can be described as plane waves, I shall first discuss plane waves.

Plane electromagnetic waves can be derived from Maxwell's equations, and it turns out that they consist of oscillating electric and magnetic fields (see Figure 2.1). At first we notice that the electric and magnetic field lines are normal to each other. Next we see that they are both in the same phase. The waves in the graph is moving from the left to the right, which we can see by applying the

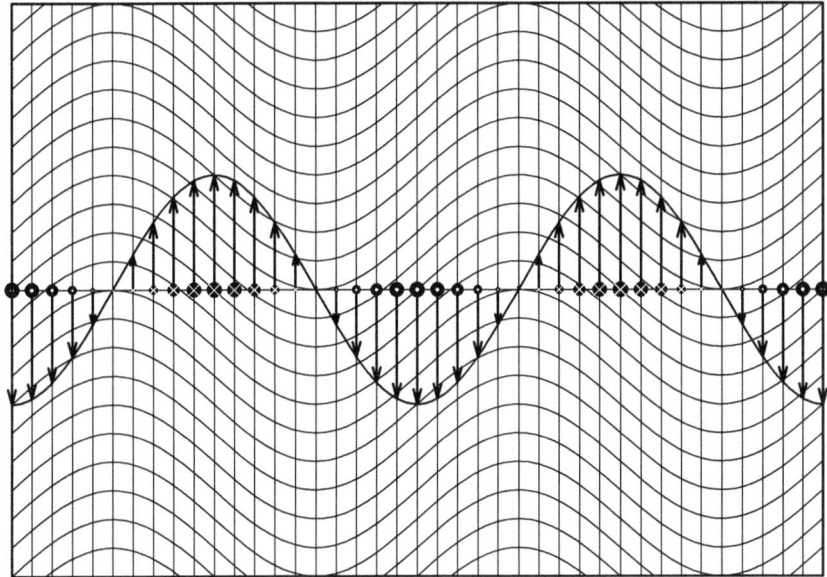

Figure 2.2: *A shear wave travelling to the right in an elastic medium (waivy black grid). Vertical arrows show the accompanying velocity field. Varying sized dots represent rotation of spatial elements (Tips and tails of arrows vertical to the velocity field).*

right hand rule. Spread the thumb, the index finger, and the middle finger on your right hand such that they are pointing in the three spatial directions. So let the index finger point in the direction of the electric field and the middle finger in the direction of the magnetic file. Then the electromagnetic wave is moving in the direction of your thumb. Apart from a couple of constants, we get the energy carried by the wave by summing up both the electric and magnetic field squared all over space. This oscillation between magnetic and electric energy also suggests that there at any time has got to be equal amounts of the two energy forms in free waves.

Next let us assume that the spatial continuum is elastic and can be deformed. Then like an ordinary material like steel, it can carry both longitudinal compression waves and transversal shear

waves. In connection with earthquakes both wave forms are known
to travel in the earth crust, and then the longitudinal waves are
called *P*-waves, for primary waves, because they travel faster and
arrive at the seismographs first. The transversal waves move with
only about half the speed of the longitudinal waves and are called
secondary waves, or *S*-waves. I shall occasionally use the same terms
for waves in the spatial continuum. They are easy to remember
since *P*-waves can be thought of as pressure waves, and *S*-waves
as shear waves because they introduce shear tensions in an elastic
continuum. And most important, the two wave forms are literally
independent of each other and can coexist in the same environment
without afflicting each other, except possibly in a non homogeneous
environment with great deformations.

In Figure 2.2 I have tried to illustrate a shear wave travelling
from left to right in the spatial continuum. The wave could have
been created by a thought experiment. Let a stiff plate normal
to the paper plane embedded in the spatial continuum be set to
rock up and down. Then a plane shear wave as illustrated, would
have been generated in both the left and right direction, but the
figure only shows the right travelling wave. As the wave moves to
the right, spatial elements alternate between moving in the up and
down direction, and can be described by a velocity field as shown by
vertical arrows. The deformation can be seen as a twist or rotation
of the spatial elements and can also be represented by arrows like
the velocity field. In the figure above they are normal to the page,
but the direction is represented by a dot to illustrate the top of
the arrow coming out of the paper, and a cross to represent the
tail of the arrow going into the page. To see the direction of the
wave movement, we again can use the right hand rule, but if we
want to let the electric field be represented by the velocity field and
the magnetic field by the rotation field, we would need to use the
negative direction of the velocity field to get it right[2].

We find that the velocity field and the rotation field relate to

---

[2]Except, of course, if we do not in stead reverse the direction of the magnetic
field, but somehow I do not think that that's likely. It does not make much of
a difference anyway, so I'll stick to the assumption above.

each other in much the same way as the electric to the magnetic field in an electromagnetic wave. That an electric field can be seen as a velocity field, and the magnetic field as a rotation, is by no way a new conception. Already in the middle of the 19th century the idea was discussed – mainly between Maxwell and Sir William Thomson (Lord Kelvin) – about the nature of electricity and magnetism. Thomson attributed as early as 1847 a linear character to electric force and electric current, and a rotatory character to magnetism, while Maxwell at first in 1855 regarded magnetic force as a linear, and electric current as a rotatory phenomenon, but he later adopted Thomson's view. In a couple of memoirs in 1861–62 [15, page 247], he wrote:

> The transference of electrolytes in fixed directions by the electric current, and the rotation of polarized light in fixed directions by magnetic force are the facts the consideration of which has induced me to regard magnetism as a phenomenon of rotation, and electric current as phenomena of translation.

Kelvin fulfilled his view in his Mathematical and Physical Papers iii, 1890, [14, 29-47, page 450-465], where he showed that in his model a linear current could be represented by a circular, or endless cord, if a tangential force were applied to the cord all around the circuit. Among others he wrote:

> ⋯. The force thus applied tangentially all round an endless line of the jelly produces a tangential drag on the jelly all around, and causes displacement and distortion ⋯ equal to half the magnetic force ⋯.

He obviously thought that the aether was highly deformable like a jelly (see Figure 2.3). In this model, however, the spatial continuum has to be many orders of magnitude harder, and the deformations correspondingly small.

So let us conclude that the electric field, apart from some constants, is like the velocity field, and the magnetic field is like the rotational field in the spatial continuum. It is then possible to show

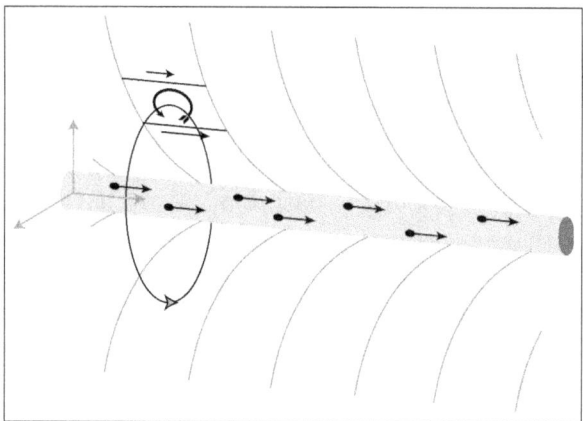

Figure 2.3: Lord Kelvin's cord. *Notice the shear forces that tend to rotate spatial elements outside the electric wire.*

that the field energy is given by the sum of the velocity field squared and the rotation field squared taken all over space just like the electromagnetic field energy[3].

If we try to see which direction the wave is moving by applying the same rule as with electromagnetic waves, we find that the thumb will point in the opposite direction of the actual wave movement. The direction of the electric field, however, was originally chosen much at random, so this discrepancy does not post any great problem. If we define the electric field as the negative velocity field in elastodynamics, then the resemblance becomes perfect. As a matter of fact, the electromagnetic field equations can be derived from the elastodynamic equations for an elastic continuum of infinite extension, even when the equations are expressed in their most precise form in a four-dimensional manifold (see Sec. 4.6, page 66).

Not only plane waves can occur in an elastic continuum. As a

---

[3]We can of course rotate a piece of steel without introducing any deformational energy into it, but in an elastic continuum of infinite extension, like the spatial continuum, a pure rotation somewhere will introduce a deformation elsewhere such that the total deformation energy always is given by summing up the rotation squared all over space (see Sec. 3.7 on page 43).

thought experiment, imagine that a small pulsating sphere is inserted somewhere in the continuum. Then a spherical longitudinal wave will be spreading uniformly in all directions with an amplitude that is decreasing inverse proportional to the distance. Moreover, solving the equation for such waves, shows that even spherical waves that are converging towards the center is possible. Also shear waves can be produced in a similar way by letting the imaginary sphere inside the body rotate clockwise and counter-clockwise around an axis through the center of the sphere in stead of pulsating. In fact, according to Huygens principle, every point of a wave front may be considered as sources of spherical waves that spread out in all directions with a speed equal to the speed of propagating waves. The superposition of all these waves will describe the total wave system at a later time.

All these are proven facts in electrodynamics and elastodynamics. We are left with a complete fit between electrodynamics and elastodynamics if we let the negative of the electric field be represented by the velocity field, and the magnetic field by the rotational field.

2.3 Maxwell's electromagnetic equations

Maxwell's equations are a set of equations, which describes all classical properties of electromagnetism. He presented his findings as eight equations in 1861, but they have later been reformulated into a modern form of four equations by Oliver Heaviside (1850–1925). (see Figure 2.4).

The two first equations are Gauss's law for magnetism and electric charge. The next is Faraday's law of induction, and the final equation is Ampère's circuital law. The first equation tells us that the magnetic field is solenoidal, i.e. the field lines do not have any beginning or end (there are no magnetic monopoles!). The next two equations states that electric fields have their sources in electric charges (ii), or may be generated by a change in the magnetic field vector (iii). The last equation, (iv), shows that the sum of the

The Heaviside version of Maxwell's equations are a set of four equations to describe all electromagnetic phenomena of nature. In SI units they take the form:

$$i) \quad \text{div } \mathbf{B} = 0,$$

$$ii) \quad \text{div } \mathbf{E} = \frac{\rho}{\varepsilon_0},$$

$$iii) \quad \text{curl } \mathbf{E} = -\frac{\partial \mathbf{B}}{\partial t},$$

$$iv) \quad \text{curl } \mathbf{B} = \mu_0 \varepsilon_0 \frac{\partial \mathbf{E}}{\partial t} + \mu_0 \, \mathbf{j}.$$

We can add the field energy equation that follows naturally from the above equations

$$v) \quad u = \frac{\varepsilon_0}{2} \, \mathbf{E}^2 + \frac{1}{2\mu_0} \, \mathbf{B}^2.$$

Figure 2.4: The Maxwell equations.

electric current and a change in the electric field vector make up a solenoidal field that creates a magnetic field around the field lines (Maxwell's correction to Ampère's law). Note also that the energy density in an electromagnetic field is given by adding up the square of the electric and magnetic field (v).

Maxwell's equations are not invariant by transformations between Cartesian coordinate systems in rectilinear movement relative to each other, e.g. a stationary charge density pattern in one coordinate system will represent a current when viewed from within a moving frame. The equations can, however, be rewritten in such a form that they are invariant by transformations between different Lorentz[4] coordinate systems. In this rather modern nomenclature, the electric and magnetic field vectors are merged together into a single tensor **F**, and the charge density and current into a vector **J** in a four-dimensional manifold. Maxwell's equations can

---

[4]Hendrik Antoon Lorentz (1853–1928)

then be written in a very compact form, and *Hermann Minkowski (1864-1909)* showed in 1908 that electromagnetic forces, the energy density, the energy transport and momentum can be expressed by the 16 components of a single stress energy tensor **T**. Written in this nomenclature the equations are invariant by transformation between different Lorentz frames that are in uniform, rectilinear movement relative to each other. On the other hand, if our measuring devises are exclusively based on electromagnetic properties, then all our reference frames would be strictly Lorentzian and it would be impossible to distinguish one Lorentzian frame from another, i.e it would be impossible to see any movements through space by considering electromagnetic phenomena, and it would fully explain why the Michelson/Morley experiment showed no movements through space. This is by the way not a far-fetched assumption since we know that all forces, except the extremely weak gravitational forces and possibly the extremely short ranged strong forces, are of electromagnetic origin. On these grounds it would be quite legible to launch the principle of relativity to all movements in space. Einstein, however, in 1905 elevated the principle of relativity to be a fundamental law of nature, and in the time that has elapsed since then, it is his view that has gained momentum, even if it still is a controversial question.

2.4 The Navier-Cauchy equation

What I am aiming at, is to get a complete description of space and matter in a model where space is an elastic continuum of infinite extension, and matter without exceptions is supposed to be disturbances and displacements in this continuum. To achieve that, I shall need a deep insight into the mechanics of solids, which I shall try to get at in a later chapter (see Chap. 3). Here I shall only outline what it is all about, and some historical facts.

In a memoir of 1821 published in 1827, *Claude Louis-Marie-Henry Navier* (1875–1836) presented an equation for an elastic solid. The equation was later developed further - mainly by *Sir George*

The Navier-Cauchy equation for an elastic continuum takes the form

$$(\lambda_s + 2\mu_s)\text{grad}\,\text{div}\,\mathbf{u} - \mu_s\text{curl}\,\text{curl}\,\mathbf{u} - \rho_s\ddot{\mathbf{u}} = -\mathbf{b}.$$

In a continuum of *infinite* extension Kelvin's theorem states that the energy density is given by

$$e = \frac{\lambda_s + 2\mu_s}{2}(\text{div}\,\mathbf{u})^2 + \frac{\mu_s}{2}(\text{curl}\,\mathbf{u})^2 + \frac{\rho_s}{2}\dot{\mathbf{u}}^2.$$

According to Helmhotz' decompositions theorem the N-C equation can be split into two literally independent equations

$$(\lambda_s + 2\mu_s)\text{grad}\,\text{div}\,\hat{\mathbf{u}} - \rho_s\ddot{\hat{u}} = -\hat{\mathbf{b}},$$
$$-\mu_s\text{curl}\,\text{curl}\,\tilde{\mathbf{u}} - \rho_s\ddot{\tilde{u}} = -\tilde{\mathbf{b}},$$

where $\mathbf{u} = \hat{\mathbf{u}} + \tilde{\mathbf{u}} = \text{grad}\,\psi + \text{curl}\,\mathbf{A}$, and $\text{div}\,\mathbf{A} = 0$.

Note that ρ_s is the spatial mass density while ρ will be reserved for the equivalence to electric charge density. Similarly the Lamé constants λ_s and μ_s are used with indices to avoid confusion with electrodynamic constants.

Figure 2.5: The Navier-Cauchy equation.

Gabriel Stokes (1819–1903) - to also include fluids, but in this paper I shall mostly consider the Linear Theory of Elasticity, which is a discipline in its own right[5]. For a homogeneous and isotropic continuum the Navier-Cauchy[6] equation takes the form shown in Fig. 2.5. The vector field, $\mathbf{u}$, represents the displacement of any points of the continuum from their original positions (tail of the arrows) to their new positions (tip of the arrows), $\dot{\mathbf{u}}$ is the speed of the translation, and $\ddot{\mathbf{u}}$ is the acceleration of a volume element, $\mathbf{b}$

---

[5]The name of the equation seems to be uncertain. It is sometimes called 'The equation of motion', or just Navier's equation. I have chosen to call it the Navier-Cauchy (or N-C) equation throughout this book.

[6]Augustin Louis Cauchy(1789-1857)

represents a (hypothetical) body force presumably from the outside word (but it can be shown to have other implications), ρ_s is the mass density, and λ_s and μ_s are Lamé's elastic constants.

The meaning of the different terms in the Navier-Cauchy equation is that of forces acting on volume elements in the elastic continuum. The term $(\lambda_s + 2\mu_s)$div $\mathbf{u}$ is a measure of the pressure in the continuum, and a varying pressure (gradient of p) will force a volume element in the direction of lower pressure. This force may be taken up by an acceleration times the mass of the volume element as the third term in the equation indicates. Further the term $\mu_s$curl$\mathbf{u}$ is a measure of rotation imposed by shear forces as suggested in Figure 2.3. The shear force naturally tends to straighten out introducing an angular momentum on the volume element which again may be taken up by dynamic forces. The outer force $\mathbf{b}$ is usually represented by gravitational or electromagnetic forces acting on volume elements in a solid, but here it is incorporated because it may be taken to describe the resulting force of a set of otherwise described forces. An electric current in a wire is supposed to exert a force of that kind, which will be discussed elsewhere.

The energy in a deformation field is given by one of Kelvin's theorems, which implies that in a body with such a great extension that there are no surface forces acting, the energy density can be expressed solely by the properties $\dot{\mathbf{u}}$, div $\mathbf{u}$, and curl $\mathbf{u}$ squared; i.e. Kelvin's theorem is valid in a continuum of infinite extension[7]. In this paper I shall only consider a spatial continuum of infinite extension (or nearly so), so Kelvin's theorem will be applicable. Notice that the energy density, e, in Box 2.5 strictly spoken is not correct unless it is summed over the entire part of the deformed space, and that no other energy, except possibly that caused by an initial uniform compression of space, is present. We could say that to a $\dot{\mathbf{u}}$, div $\mathbf{u}$, and curl $\mathbf{u}$ there always corresponds an energy given by e somewhere in space. Thus we take the liberty to call e the energy density just like what is common in electrodynamics. For a more

---

[7]Since matter is noting but disturbance energy in the elastic continuum, this is a first indication that matter is none local. The energy equivalent to mass of a body can in reality be located anywhere in space.

thorough discussion, see Sec. 3.7.

Hermann von Helmholtz (1821–94) has shown that a vector field can be divided into an irrotational (curl-free) and an solenoidal (divergence-free) field. The field lines in an irrotational field will tend to originate from a source and never run in a closed loop while the field lines in a solenoidal field always will run in closed loops and never pop out from a source. These two fields give rise to different wave equations describing longitudinal and transversal waves respectively, the former moving with about the double speed of the latter. These two waveforms are literally independent of each other, and even if they are initiated at the same point (as they might be for instance by an earthquake) they live their own lives and spread independently (e.g. as P and S waves). Thus in all dynamic connections the Navier-Cauchy equation can be divided into one *irrotational* and one *solenoidal* part. Only when we are dealing with fields near a singularity will the Navier-Cauchy equation break down and the two fields are no longer independent of each other.

So much for history. In the next section I shall try to compare the N-C equation and some mathematical identities in an isotropic and homogeneous elastic continuum of (nearly) infinite extension with Maxwell's equations. I shall do a more detailed comparison in Chapter 4 where also the relativistic properties will be discussed.

2.5   Maxwell's and Navier-Cauchy's equation

In this section I shall make a comparison between Maxwell's equations and the solenoidal part of the Navier-Cauchy equation for an elastic continuum of infinite extension. By redefining the terms in the N-C equation slightly, we get a new set of equations that formally are like Maxwell's equations except from equation ii) that needs a closer examination (see Figure 2.6).

In the discipline of hydrodynamics it is common to use hypothetical sinks and sources as means to mathematically describe real moving bodies of different shapes in a perfect fluid. Mass disappears

By reformulating the terms in the N-C equation

$$\mathbf{b} = \mathbf{j}, \quad \dot{\mathbf{u}} = -\mathbf{E}, \quad \text{curl } \mathbf{u} = \mathbf{B}, \quad \rho_s = \varepsilon_0, \quad \mu_s = \frac{1}{\mu_0},$$

it transforms into

$$iv) \qquad \text{curl } \mathbf{B} = \varepsilon_0 \mu_0 \frac{\partial \mathbf{E}}{\partial t} + \mu_0 \mathbf{j}.$$

Generally we have that the divergence of a curl is zero, so

$$i) \qquad \text{div } \mathbf{B} = 0,$$

and by the identity $\frac{\partial (\text{curl } \mathbf{u})}{\partial t} = \text{curl } \left(\frac{\partial \mathbf{u}}{\partial t} \right)$ we have

$$iii) \qquad \frac{\partial \mathbf{B}}{\partial t} = -\text{curl } \mathbf{E}.$$

The density of sinks can be written as

$$ii) \qquad \text{div } \mathbf{E} = \frac{\rho}{\varepsilon_0}$$

By Kelvin's theorem the field energy density becomes

$$v) \qquad e = \frac{\varepsilon_0}{2} \mathbf{E}^2 + \frac{1}{2\mu_0} \mathbf{B}^2.$$

Figure 2.6: Reformulating the N-C equation.

into a sink and reappear out of a source, almost as there should be a secret loophole between them. Here, however, I shall assume that there may be free sinks and sources as real point-like entities in the spatial continuum. How such entities can be realized will be discussed elsewhere, but here it is only necessary to state that a source will be seen as a negative sink, and that they can only be created in pairs, one equally strong source for every sink. Let the strength of a sink be defined as the inflow of spatial mass per time unit, and let a number of sinks be smoothly distributed in space.

We can then define a *sink density*, ρ, as the sum of all sinks in a small volume element – that still is great enough to contain many sinks – divided by the volume of the volume element. We can then put up the last equation that completes the comparison between the Navier-Cauchy equation and Maxwell's equations. We also find that there is a hidden dependency between the properties ρ and $\mathbf{j}$ (see Fig. 2.6 and 2.7). If our initial condition holds that there may be freely movable sinks and sources in the spatial continuum, and that they can only be created by pair production, then Equation (vi) in Fig. 2.7 can only be interpreted as a continuity equation, meaning that a change of sink density in an area can only be accomplished by an in- or outflow of sinks. Hence $\mathbf{j}$ has got to represent a flow of sinks, and further, since $\mathbf{j}$ originally was set like the body force $\mathbf{b}$, that a flow of sinks will create a drag in the spatial continuum just as Lord Kelvin's proposed in 1890 (see Figure 2.3).

The assumption that div $\dot{\mathbf{u}} \neq 0$ while div $\mathbf{u} \equiv 0$ can only be realized by assuming that the equivalents to positive and negative electric charges are the point-like entities, sinks and sources respectively. Thus everywhere in between the sinks and sources, and hence all over space since the singularities themselves do not contribute to the mean density of the spatial continuum, we have that div $\mathbf{u} = 0$ as required. We know from Quantum Electro Dynamics (QED) that it is possible to explain forces between electric charges by an exchange of photons between them, so a further discussion about how sinks and sources can be formed in a continuum, has got to be discussed in connection with the nature of electric charges, but once sinks and sources are realized as viable entities, they will behave exactly like electric charges and exert the drag on the spatial continuum that Lord Kelvin predicted.

The above equations were all developed in a Euclidean coordinate system, and they are not invariant by transformation between such coordinate systems in relative motion to each other. For example can a static pattern of sinks and sources in one frame be seen as a flow of the entities in another. It is, however, possible to go a step further and show that the fields may be developed in frame independent form, which make them invariant by transformations

Let q_τ be the sum of all sinks (sources are negative sinks) inside a volume element τ, and let τ shrink towards a little volume ϵ that still contains many sources. Then we can define a sink density given by

$$\rho = \lim_{\tau \to \epsilon} \tfrac{q_\tau}{\tau} = -\varepsilon_0 \lim_{\tau \to \epsilon} \tfrac{1}{\tau} \oint_\tau \dot{\mathbf{u}} \, \mathbf{n} \, df = -\varepsilon_0 \mathrm{div} \, \dot{\mathbf{u}},$$

$$ii) \quad \rho \qquad\qquad\qquad = \varepsilon_0 \mathrm{div} \, \mathbf{E}.$$

There is a dependency between ρ and $\mathbf{j}$.
Take the divergence of Equation iv):

$$\mathrm{div} \, \mathrm{curl} \, \mathbf{B} = \varepsilon_0 \mu_0 \mathrm{div} \, \dot{\mathbf{E}} + \mu_0 \mathrm{div} \, \mathbf{j},$$

$$\varepsilon_0 \mathrm{div} \, \dot{\mathbf{E}} = -\mathrm{div} \, \mathbf{j},$$

and take the partial derivative of Equation ii) with respect on t:

$$\varepsilon_0 \mathrm{div} \, \dot{\mathbf{E}} = \dot{\rho}.$$

By subtracting the two equations from each other we obtain the continuity equation

$$vi) \quad \dot{\rho} + \mathrm{div} \, \mathbf{j} = 0.$$

Figure 2.7: Density of sinks

between different Lorentz coordinate systems. Written in this form the equations are *coordinate invariant*, – i.e. if the equations hold in one coordinate system, they will hold in any. Hence, even if the initial comparison between elastodynamic and electromagnetic fields were performed in a fixed frame, the result would be equally valid in any Lorenz frame in uniform rectilinear motion. I'll return to this question in Chap. 4.

Naturally the question of which frame the observer measures the phenomena in, will immediately arise. The observer has got to rely on measuring rods and clocks that she brings with her, and if they are subject to changes when they move along with her as she

performs her observations, the result will be affected. Moreover, if
the devices for measuring length and time is changed in agreement
with the Lorenz contraction and time dilatation as *G.F. Fitzgerald
(1851-1901)* did propose, then the observer's frame would be strictly
Lorentzian and she would have no means whatsoever to find out how
she is moving through the spatial continuum.[8]

This leads us to a philosophical question. If we cannot detect
any motion through space with any of our means even if there really
is a fixed reference frame, then why bother about it? We should use
Occam's razor and dispose of the whole concept: *There is no ether
out there!* This is what was done at the passage from the 19th to
the 20th century, and the result was Special Relativity. Since then
there has never been discovered any phenomenon that enables us
to measure a speed through space, so Special Relativity has proved
very well fitted to describing all kinds of movements explicitly in
relation to other bodies, because a discrepancy would mean that an
Ether would be detectable. The principle, which has been ascribed
to *William of Occam (c. 1295 – 1349)*, is usually translated from
Latin to mean that *entities should not be multiplied beyond necessity.* It states that one should not take into account more than is
necessary to describe a phenomenon, hence exit of the Ether. Even
if it should be possible to detect the spatial continuum with some
very subtle experiments in the future, perhaps by detecting a difference in wave speed between light and gravitational waves, Special
Relativity will always be a handy tool to describing the bulk of
observable phenomenon.

---

[8]From Larmor, Joseph (1900), Aether and matter, Cambridge, [England]:
Cambridge University Press : n.p. I quote:

> ... if the internal forces of a material system arise wholly from elec-
> tromagnetic actions between the system of electrons which consti-
> tute the atoms, then the effect of imparting to a steady material
> system a uniform velocity of translation is to produce a uniform
> contraction of the system in the direction of motion, of amount
> $\sqrt{1 - v^2/c^2}$.

Chapter 3

Elastodynamics

The *Linear Theory of Elasticity* is a discipline in its own right. The theory was probably originally intended to describe ordinary elastic bodies consisting of particles bound together by molecular forces, but it is through the centuries refined to be a theory that can describe deformations in a true *elastic continuum*. Here, I will focus on the part of the theory that describes deformations in an elastic continuum of *infinite extension* or nearly so. In its undeformed state it is supposed to be *unbounded, homogeneous,* and *isotropic*. A major part of the theory of elasticity is about surface forces on bodies of different shapes, crystalline anisotropy, inhomogeneities, and so on, but none of such themes will be discussed here. For more reading, I refer to *Mechanics of Solids II*, Volume VIa/2 of *Encyclopedia of Physics*. Springer, 1972.

3.1 Displacement fields

The space B under consideration is an *elastic continuum of infinite extension*, which is *homogeneous* and *isotropic* and has a *mass density* of ρ_s[1]. It obeys the deformation laws of The Linear Theory

---

[1]Notice that I already now put the index s on the mass density, i.e. $\rho = \rho_s$, and somewhat later, also on the shear modulus, i.e. $\mu = \mu_s$, in order to distinguish them from the charge density, ρ, and the the permeability, μ, in

of Elasticity for deformations that are so small that they can be considered infinitesimal.

The displacement in this space is described by the *displacement field*; its value $\mathbf{u}(\mathbf{x}, t)$ at a *point* $\mathbf{x}$ is the 'infinitesimal' *displacement* of a point P from $\mathbf{x}$ to $\mathbf{X}$. The symmetric part of the *displacement gradient*, $\nabla\mathbf{u}$, given by

$$\epsilon \;=\; \tfrac{1}{2}\left(\nabla\mathbf{u} + \nabla\mathbf{u}^T\right), \tag{3.1}$$

is the infinitesimal *strain field*, and the above equation, relating ϵ to $\mathbf{u}$, is called the *strain-displacement relation*.

By introducing the coordinates, $x_1 = x$, $x_2 = y$, $x_3 = z$, and letting the Roman indices run from 1 to 3, this relation can be written i different coordinate forms:

$$\epsilon_{ij} \;=\; \tfrac{1}{2}\left(\frac{\partial u_i}{\partial x_j} + \frac{\partial u_j}{\partial x_i}\right),$$
$$\epsilon_{ij} \;=\; \tfrac{1}{2}\left(\partial_j u_i + \partial_i u_j\right),$$
$$\epsilon_{ij} \;=\; \tfrac{1}{2}\left(u_{i,j} + u_{j,i}\right).$$

The tensor, ϵ_{ij}, can also be written as a matrix:

$$\epsilon_{ij} = \begin{bmatrix} \epsilon_{11} & \epsilon_{12} & \epsilon_{13} \\ \epsilon_{21} & \epsilon_{22} & \epsilon_{23} \\ \epsilon_{31} & \epsilon_{32} & \epsilon_{33} \end{bmatrix}.$$

These notations will be used frequently throughout this paper.

All deformations will take place in a *local space*, which in this context is to be understood as the whole of the deformed area, or at least an area through which border no significant forces due to the inside deformation are conveyed. In addition $\mathbf{u}$ has got to be continuous and sufficiently smooth, i.e. the spatial continuum will never be fractured.

We call

$$\operatorname{div}\mathbf{u} \;=\; \operatorname{tr}\epsilon,$$
$$u_{j,j} \;=\; \delta_i{}^j \epsilon_{ij},$$

---

electrodynamics.

the *dilatation*. Here the Euclid metric, $\delta_i{}^j$, is given by

$$\delta_i{}^j = \begin{bmatrix} 1\,0\,0 \\ 0\,1\,0 \\ 0\,0\,1 \end{bmatrix},$$

and the Einstein summation convention, by which like indices are summed over, is applied throughout the rest of this book.

The displacement field, $\mathbf{u}$, is said to be *solenoidal* if $\operatorname{div}\mathbf{u} \equiv 0$, and *irrotational* if $\operatorname{curl}\mathbf{u} \equiv 0$ all over space.

3.2 System of forces

In an elastic continuum there may be a *system of forces* acting on a part, B, of space bounded by a surface, ∂B. The resulting force on B is the sum of three possible forces: A body force, the sum of surface forces, and a counter-forces caused by acceleration.

The most familiar example of a body force is gravity, which is not present in the spatial continuum. I shall all the same keep the possibility open for the occurrence of a hypothetical *body force*, $\mathbf{b}$.

The continuum can be subject to stresses, which can be described by a stress tensor, σ. The force acting on a surface element of ∂B is given by $\Delta\mathbf{f} = \sigma \cdot \mathbf{n}\,\Delta a$, where Δa is a surface element of ∂B, and $\mathbf{n}$ is a unit vector normal to Δa. The *surface force* acting on the whole surface, ∂B, i.e. the body B, is given by

$$\mathbf{F}_B = \oint_{\partial B} \sigma \cdot \mathbf{n}\,da,$$

and per unit volume it amounts to

$$\mathbf{f} = \lim_{V_B \to 0} \frac{1}{V_B} \oint_{\partial B} \sigma \cdot \mathbf{n}\,da.$$

This is simply the divergence of the tensor, σ, and we obtain

$$\mathbf{f} = \operatorname{div}\sigma. \tag{3.2}$$

We assume that there all over space is a strictly positive function ρ_s called the *density* such that the mass of B is given by

$$m_B = \int_B \rho_s \, dV$$

The *motion* of the body is described by the (infinitesimal) *displacement field* $\mathbf{u}(\mathbf{x}, t)$ such that

$$\dot{\mathbf{u}} = \frac{\partial \mathbf{u}}{\partial t} \quad \text{and} \quad \ddot{\mathbf{u}} = \frac{\partial^2 \mathbf{u}}{\partial^2 t}$$

are the *velocity* and *acceleration* of B respectively. The *linear momentum* $\mathbf{l}$ of B is

$$\mathbf{l}_B = \int_B \rho_s \dot{\mathbf{u}} \, dV,$$

and the counterforce $\mathbf{F}$ on B caused by acceleration is

$$\mathbf{F}_B = - \int_B \rho_s \ddot{\mathbf{u}} \, dV.$$

Per unit volume it amounts to

$$\mathbf{f}_B = -\rho_s \ddot{\mathbf{u}},$$

minus because it is a counter-force.

This set of forces has got to sum up to zero, and we obtain:

$$\operatorname{div} \sigma + \mathbf{b} - \rho_s \ddot{\mathbf{u}} = 0.$$

The *Cauchy-Poisson theorem* [6, page 44] states that if $\mathbf{u}$ is an admissible motion and f is a system of forces, then $[\mathbf{u}, f]$ is a dynamic process if and only if the following two conditions are satisfied:

1. there exists a symmetric tensor field σ called the *stress field*, such that for each unit vector $\mathbf{n}$, the force on a surface normal to $\mathbf{n}$ is

$$\mathbf{f}_n = \sigma \cdot \mathbf{n} \, ;$$

2. $\mathbf{u}, \sigma$, and $\mathbf{b}$ satisfy the *equation of motion*

$$\operatorname{div} \sigma + \mathbf{b} = \rho_s \, \ddot{\mathbf{u}} \,. \tag{3.3}$$

This theorem is one of the major results of continuum mechanics.

3.3 The stress-strain relation

In a linearly elastic continuum there exists a relation between strain and the stress it causes, which can be expressed by the relation

$$\sigma(\mathbf{x}) = \mathbf{C}\big[\epsilon(\mathbf{x})\big],$$

where $\mathbf{C}$ is a fourth-order symmetric tensor that maps the space of strain onto the space of stress according to

$$\sigma_{ij} = C_{ijkl}\epsilon_{kl} \,.$$

$\mathbf{C}$ is called the *elasticity tensor*. The spatial continuum under consideration is both homogeneous and isotropic and these properties immediately reduces the 81 components of $\mathbf{C}$ such that $\mathbf{C}$ may be described by only two different scalar constants. The stress-strain relation in a homogeneous and isotropic continuum thus takes the relatively simple form

$$\begin{aligned} \sigma_{ij} &= 2\mu_s\epsilon_{ij} + \lambda_s\epsilon_{kk}\delta_{ij} \,, \\ &= \mu_s(u_{i,j} + u_{j,i}) + \lambda_s u_{k,k}\delta_{ij} \,, \end{aligned} \tag{3.4}$$

where μ_s and λ_s are *Lamé's elastic moduli*[2], which are constants in a homogeneous elastic continuum, and δ_{ij} is the *Kronecker delta*

$$\delta_{ij} = \begin{cases} 1 & \text{if} \quad i = j \,, \\ 0 & \text{if} \quad i \neq j \,. \end{cases}$$

---

[2]Note that I have put on the indices s to avoid mixing them up with other properties in electrodynamics.

3.4   The Navier-Cauchy equation

From the strain field (3.1), the Stress-strain relation (3.4) and the equation of motion (3.3) one can derive the *Navier-Cauchy equation* [6, page 213][3]

$$
\begin{aligned}
\mu_s u_{i,jj} + \mu_s u_{j,ij} + (\lambda_s u_{k,k}\delta_{ij})_{,j} + b_i &= \rho_s \ddot{u}_i \\
\mu_s u_{i,jj} + \mu_s u_{j,ji} + \lambda_s u_{k,ki} + b_i &= \rho_s \ddot{u}_i, \\
\mu_s \nabla^2 \mathbf{u} + (\lambda_s + \mu_s)\nabla \operatorname{div} \mathbf{u} + \mathbf{b} &= \rho_s \ddot{\mathbf{u}},
\end{aligned}
\tag{3.5}
$$

or equivalently by the mathematical identity

$$
\operatorname{curl}\operatorname{curl}\mathbf{A} = \nabla \operatorname{div}\mathbf{A} - \nabla^2 \mathbf{A},
\tag{3.6}
$$

we obtain

$$
\begin{aligned}
(\lambda_s + 2\mu_s)\nabla \operatorname{div}\mathbf{u} - \mu_s \operatorname{curl}\operatorname{curl}\mathbf{u} + \mathbf{b} &= \rho_s \ddot{\mathbf{u}}, \\
\frac{\lambda_s + 2\mu_s}{\rho_s}\operatorname{grad}\operatorname{div}\mathbf{u} - \frac{\mu_s}{\rho_s}\operatorname{curl}\operatorname{curl}\mathbf{u} + \frac{\mathbf{b}}{\rho_s} &= \ddot{\mathbf{u}}.
\end{aligned}
$$

By defining two new constants[4]

$$
c_g = \sqrt{\frac{\lambda_s + 2\mu_s}{\rho_s}}, \quad c_l = \sqrt{\frac{\mu_s}{\rho_s}},
$$

the Navier-Cauchy equation takes the form

$$
c_g{}^2 \operatorname{grad}\operatorname{div}\mathbf{u} - c_l{}^2 \operatorname{curl}\operatorname{curl}\mathbf{u} + \frac{\mathbf{b}}{\rho_s} = \ddot{\mathbf{u}}.
\tag{3.7}
$$

---

[3] Also termed the *Navier equations of motion* for a homogeneous and isotropic linear elastic solid [12].

[4] In this model of space and matter c_g probably turns out to be the speed of gravity while c_l is the speed of light. Hence the choice of indexes.

3.5 Wave movements

Generally we have that by taking the divergence of curl $\mathbf{A}$, and the curl of div $\mathbf{B}$, where $\mathbf{A}$ and $\mathbf{B}$ are arbitrary vectors, the results in both cases are zero. So by operating on the Navier-Cauchy equation first with the *divergence* operator and next with the *curl* operator, we obtain respectively:

$$\nabla^2(\mathrm{div}\,\mathbf{u}) - \frac{1}{c_g^{\,2}}\frac{\partial^2(\mathrm{div}\,\mathbf{u})}{\partial t^2} = -\frac{\mathrm{div}\,\mathbf{b}}{\lambda_s + 2\mu_s},$$

$$\nabla^2(\mathrm{curl}\,\mathbf{u}) - \frac{1}{c_l^{\,2}}\frac{\partial^2(\mathrm{curl}\,\mathbf{u})}{\partial t^2} = -\frac{\mathrm{curl}\,\mathbf{b}}{\mu_s}.$$

With $\mathbf{b} = \mathbf{0}$ we have two wave equations where the dilatation, div $\mathbf{u}$, satisfies a wave moving with the speed c_g, while the rotational component, curl $\mathbf{u}$, satisfies a wave moving with the speed c_l. In fact the *Propagation theorem for isotropic bodies* states that if a body is isotropic, then a wave is either *longitudinal*, in which case $c = c_g$, or *transversal*, in which case $c = c_l$ [6, page 256]. By the way, such waves are known in connection with earthquakes, which normally produces both wave forms. The longitudinal waves are called P-waves, or primary waves, since they approach at the seismometers first, and the transversal waves are for the same reason called S-waves, or secondary waves.

3.6 Solenoidal and irrotational deformations

According to *Helmholtz's Theorem*, any vector field that satisfies the condition

$$(\nabla \cdot \mathbf{a})_\infty = 0,$$

$$(\nabla \times \mathbf{a})_\infty = 0,$$

(i.e. $\mathbf{a} \equiv \mathbf{0}$ at an infinite distance from the considered area) may be written as the sum of an irrotational part and a solenoidal part,

$$\mathbf{a} = -\nabla\phi + \nabla \times \mathbf{A}, \qquad\qquad (3.8)$$

Hence the displacement field can be decomposed into two properties

$$\mathbf{u} = \hat{\mathbf{u}} + \tilde{\mathbf{u}},$$

where

$$
\begin{aligned}
\hat{\mathbf{u}} &= -\operatorname{grad}\phi, \\
\tilde{\mathbf{u}} &= \operatorname{curl}\psi, \qquad \operatorname{div}\psi = 0.
\end{aligned}
$$

So, because the divergence of a curl, and the curl of a divergence always are zero, we obtain:

$$
\begin{aligned}
\operatorname{div}\mathbf{u} &= \operatorname{div}\hat{\mathbf{u}}, \\
\operatorname{curl}\mathbf{u} &= \operatorname{curl}\tilde{\mathbf{u}}.
\end{aligned}
$$

The same aplies of course to $\mathbf{b}$, so the two wave equations in the previous section can be written:

$$
\begin{aligned}
\nabla^2(\operatorname{div}\hat{\mathbf{u}}) - \frac{1}{c_g^{\,2}}\frac{\partial^2(\operatorname{div}\hat{\mathbf{u}})}{\partial t^2} &= -\frac{\operatorname{div}\hat{\mathbf{b}}}{\lambda_s + 2\mu_s}, \\
\nabla^2(\operatorname{curl}\tilde{\mathbf{u}}) - \frac{1}{c_l^{\,2}}\frac{\partial^2(\operatorname{curl}\tilde{\mathbf{u}})}{\partial t^2} &= -\frac{\operatorname{curl}\tilde{\mathbf{b}}}{\mu_s}.
\end{aligned}
$$

By the mathematical identity, $\operatorname{curl}\operatorname{curl}\mathbf{A} = \operatorname{grad}\operatorname{div}\mathbf{A} - \nabla^2\mathbf{A}$, we have:

$$
\begin{aligned}
\nabla^2(\operatorname{div}\hat{\mathbf{u}}) &= \operatorname{grad}\operatorname{div}(\operatorname{div}\hat{\mathbf{u}}) - \operatorname{curl}\operatorname{curl}(\operatorname{div}\hat{\mathbf{u}}) \\
&= \operatorname{grad}\operatorname{div}(\operatorname{div}\hat{\mathbf{u}}),
\end{aligned}
$$

and

$$
\begin{aligned}
\nabla^2(\operatorname{curl}\tilde{\mathbf{u}}) &= \operatorname{grad}\operatorname{div}(\operatorname{curl}\tilde{\mathbf{u}}) - \operatorname{curl}\operatorname{curl}(\operatorname{curl}\tilde{\mathbf{u}}) \\
&= -\operatorname{curl}\operatorname{curl}(\operatorname{curl}\tilde{\mathbf{u}}).
\end{aligned}
$$

By considering the N-C equation in *a spatial continuum of infinite extension,* we have obtained two different equations: One for *irrotational deformations*:

$$\operatorname{grad}\operatorname{div}\hat{\mathbf{u}} + \frac{\hat{\mathbf{b}}}{\lambda_s + 2\mu_s} = \frac{1}{c_g^{\,2}}\ddot{\hat{\mathbf{u}}} \tag{3.9}$$

and one for *solenoidal deformations*:

$$-\operatorname{curl}\operatorname{curl}\tilde{\mathbf{u}} + \frac{\tilde{\mathbf{b}}}{\mu_s} = \frac{1}{c_l{}^2}\ddot{\tilde{\mathbf{u}}} \tag{3.10}$$

The division of the Navier-Cauchy equation into one irrotational and one solenoidal part, allows us to examine these two parts separately and thereby simplify the strain-stress relations immensely. Very much in the rest of this book will imply displacements, so I'll distinguish between $\hat{u}$ and $\tilde{u}$ only when it is necessary for clarity. Mostly I'll only use the unmarked symbol throughout.

3.7 Field energy in the spatial continuum

From the Navier-Cauchy equation one can find the internal field energy in an admissible field in B by performing the following thought experiment: Introduce a hypothetical body force, $-\mathbf{b}$ (negative because $\mathbf{b}$ is a breaking force), from the outside world such that it eradicates the entire field in B; i.e. $\mathbf{u}$ and all functions of $\mathbf{u}$ become constant like zero all over B. In addition I will assume that the entire field is confined inside B such that $\mathbf{u}$ is zero on the surface of B and beyond. The energy released by this operation, E, would then be like the total field energy in B:

$$
\begin{aligned}
E &= -\int_B dv \int_{f(\mathbf{u})}^{0} \mathbf{b}\,d\mathbf{u} \\
&= \int_B dv \left[\int_{0}^{f(\mathbf{u})} (\rho_s\ddot{\mathbf{u}} - (\lambda_s + 2\mu_s)\operatorname{grad}\operatorname{div}\mathbf{u} + \mu_s\operatorname{curl}\operatorname{curl}\mathbf{u})\,d\mathbf{u} \right]
\end{aligned}
$$

E can be separated into three integrals, i.e. $E = E_1 + E_2 + E_3$. The first of these integrals is simply the kinetic energy of the system

$$E_1 \;=\; \int_B dv \left(\rho_s \int_0^{\dot{\mathbf{u}}} \frac{d\dot{\mathbf{u}}}{dt} d\mathbf{u} \right) = \int_B dv \left(\rho_s \int_0^{\dot{\mathbf{u}}} d\dot{\mathbf{u}} \cdot \dot{\mathbf{u}} \right),$$

$$E_1 \;=\; \int_B \frac{1}{2} \rho_s \dot{\mathbf{u}}^2 dv.$$

The next part can be integrated by using the mathematical identity:

$$\operatorname{div}(\phi \mathbf{A}) = \phi \operatorname{div} \mathbf{A} + \mathbf{A} \operatorname{grad} \phi, \tag{3.11}$$

and inserting $\phi = \operatorname{div} \mathbf{u}$ and $\mathbf{A} = d\mathbf{u}$

$$E_2 \;=\; (\lambda_s + 2\mu_s) \int_B dv \int_0^{\operatorname{div}\mathbf{u}} [\operatorname{div}\mathbf{u} \cdot \operatorname{div}(d\mathbf{u}) - \operatorname{div}(d\mathbf{u} \cdot \operatorname{div}\mathbf{u})]$$

$$= (\lambda_s + 2\mu_s) \int_B dv \int_0^{\operatorname{div}\mathbf{u}} [\operatorname{div}\mathbf{u} \cdot d(\operatorname{div}\mathbf{u})]$$

$$- (\lambda_s + 2\mu_s) \int_B dv \cdot \operatorname{div}\left(\int_0^{\operatorname{div}\mathbf{u}} d\mathbf{u} \cdot \operatorname{div}\mathbf{u} \right).$$

The first part of the integral can readily be integrated, and the last part can be transformed into a surface integral over ∂B by the Divergence theorem[5] and disappear because $\mathbf{u}$ is constant like zero on the border of B and beyond. Thus

$$E_2 = \int_B \frac{1}{2} (\lambda_s + 2\mu_s)(\operatorname{div}\mathbf{u})^2 dv.$$

---

[5] $\oint_{\partial B} (\mathbf{A} \cdot \mathbf{n}) \, df = \int_B \operatorname{div} \mathbf{A} \, dv$

We can find E_3 in much the same way by using the identity

$$\text{div} (\mathbf{A} \times \mathbf{B}) = \text{curl}\, \mathbf{A} \cdot \mathbf{B} - \text{curl}\, \mathbf{B} \cdot \mathbf{A}, \tag{3.12}$$

and inserting $\mathbf{B} = \text{curl}\, \mathbf{u}$ and $\mathbf{A} = d\mathbf{u}$

$$E_3 = -\mu_s \int_B dv \int_0^{\text{curl}\, \mathbf{u}} \left[\text{div} (d\mathbf{u} \times \text{curl}\, \mathbf{u}) - \text{curl}\, \mathbf{u} \cdot \text{curl} (d\mathbf{u}) \right]$$

$$= \mu_s \int_B dv \int_0^{\text{curl}\, \mathbf{u}} \text{curl}\, \mathbf{u} \cdot d(\text{curl}\, \mathbf{u})$$

$$-\mu_s \int_B dv\, \text{div} \left(\int_0^{\text{curl}\, \mathbf{u}} d\mathbf{u} \times \text{curl}\, \mathbf{u} \right).$$

Again the first part can be integrated and the last part disappear by the same reason as above, and we get

$$E_3 = \int_B \frac{1}{2} \mu_s (\text{curl}\, \mathbf{u})^2 dv.$$

Finally we can write the total energy in the deformed area:

$$E = \frac{1}{2} \int_B \left[\rho_s \dot{\mathbf{u}}^2 + (\lambda_s + 2\mu_s)(\text{div}\, \mathbf{u})^2 + \mu_s (\text{curl}\, \mathbf{u})^2 \right] dv. \tag{3.13}$$

The result is already known as *Kelvin's theorem* [6, page 208], and is a proven theorem in the Linear theory of elasticity. The result can be interpreted as the *local energy density* even if this development does not say anything about where in the field the energy is to be found; only that there to a $\dot{\mathbf{u}}$, $\text{curl}\, \mathbf{u}$, and $\text{div}\, \mathbf{u}$ always corresponds an energy given by the equation above, and no other energy is present as long as we deal with infinitesimal deformations restricted to a limited area of a homogeneous and isotropic continuum covered by the Linear theory of elasticity.[6] We often enough

---

[6]The corresponding expression for the energy density in an electromagnetic field has the same limitation, but nonetheless it is usually interpreted as the local energy density.

has got to speak about energy and even mass density, but we should very carefully bear in mind that energy density only gives meaning when integrated over the whole of space. A mathematical expression only tells us that space acts "*as if*" there is a certain local energy density in the region. It is pretty obvious that if we should happen to find a seemingly point like entity with a huge amount of energy, then the energy cannot be located inside that point. This is the first clue that the mathematics does not always give us the full description of the physical state of a system.

With this restriction in mind the local energy density, e, in the spatial continuum is given by

$$e = \frac{1}{2}\rho_s \dot{\mathbf{u}}^2 + \frac{1}{2}(\lambda_s + 2\mu_s)(\text{div }\mathbf{u})^2 + \frac{1}{2}\mu_s(\text{curl }\mathbf{u})^2. \qquad (3.14)$$

Because I did not state anything about a possible initial pressure in the spatial continuum, only that $\mathbf{u}$ at the border of B is not affected by deformations within B, we notice that any additional deformation in B produces a positive energy. It is noteworthy that the energy density in any field of strain and motion is non-negative even if the space itself should happen to contain a huge amount of uniformly distributed energy due to an initial pressure.

3.8   Energy transport

The *energy transport* in the deformation field can be found by deriving the energy equation with respect on time. We acquire

$$\frac{\partial e}{\partial t} = \rho_s\,\dot{\mathbf{u}}\,\ddot{\mathbf{u}} + (\lambda_s + 2\mu_s)\text{div }\mathbf{u}\,\text{div }\dot{\mathbf{u}} + \mu_s\text{curl }\mathbf{u}\,\text{curl }\dot{\mathbf{u}}.$$

We substitute $\rho_s\ddot{\mathbf{u}}$ from the N-C Equation, (3.7), and get

$$\begin{aligned}
\frac{\partial e}{\partial t} = &\ \dot{\mathbf{u}}\big[(\lambda_s + 2\mu_s)\text{grad div }\mathbf{u} - \mu_s\text{curl curl }\mathbf{u} + \mathbf{b}\big] \\
&+ (\lambda_s + 2\mu_s)\text{div }\mathbf{u}\cdot\text{div }\dot{\mathbf{u}} + \mu_s\text{curl }\mathbf{u}\cdot\text{curl }\dot{\mathbf{u}}, \\
\mathbf{b}\dot{\mathbf{u}} = &\ \frac{\partial e}{\partial t} + \mu_s\big[\text{curl }(\text{curl }\mathbf{u})\cdot\dot{\mathbf{u}} - \text{curl }\dot{\mathbf{u}}(\text{curl }\mathbf{u})\big] \\
&- (\lambda_s + 2\mu_s)\big[(\text{div }\mathbf{u})\text{div }\dot{\mathbf{u}} + \dot{\mathbf{u}}\text{grad }(\text{div }\mathbf{u})\big]. \qquad (3.15)
\end{aligned}$$

By the identity below and setting $\mathbf{A} = \text{curl}\,\mathbf{u}$, and $\mathbf{B} = \dot{\mathbf{u}}$, the third term in (3.15) becomes:

$$\text{curl}\,\mathbf{A} \cdot \mathbf{B} - \text{curl}\,\mathbf{B} \cdot \mathbf{A} \;=\; \text{div}\,(\mathbf{A} \times \mathbf{B}),$$
$$\text{curl}\,(\text{curl}\,\mathbf{u}) \cdot \dot{\mathbf{u}} - \text{curl}\,\dot{\mathbf{u}}(\text{curl}\,\mathbf{u}) \;=\; \text{div}\,(\text{curl}\,\mathbf{u} \times \dot{\mathbf{u}}).$$

Further, by the identity, $\text{div}\,(\phi\mathbf{A}) = \phi\,\text{div}\,\mathbf{A} + \mathbf{A}\,\text{grad}\,\phi$, and setting $\phi = \text{div}\,\mathbf{u}$, and $\mathbf{A} = \dot{\mathbf{u}}$, the last term in (3.15) becomes:

$$\text{div}\,\mathbf{u} \cdot \text{div}\,\dot{\mathbf{u}} + \dot{\mathbf{u}}\,\text{grad}\,(\text{div}\,\mathbf{u}) \;=\; \text{div}\,(\text{div}\,\mathbf{u} \cdot \dot{\mathbf{u}}).$$

With these results inserted, (3.15) ends up as:

$$\frac{\partial e}{\partial t} + \text{div}\,(\mu_s \text{curl}\,\mathbf{u} \times \dot{\mathbf{u}}) - \text{div}\left[(\lambda_s + 2\mu_s)\text{div}\,\mathbf{u} \cdot \dot{\mathbf{u}}\right] = \mathbf{b}\dot{\mathbf{u}}.$$

Now we define two new vectors

$$
\begin{aligned}
\hat{\mathbf{S}} &= -(\lambda_s + 2\mu_s)\dot{\mathbf{u}} \cdot \text{div}\,\mathbf{u}\,], \\
\tilde{\mathbf{S}} &= -\mu_s \dot{\mathbf{u}} \times \text{curl}\,\mathbf{u}\,,
\end{aligned}
\tag{3.16}
$$

and obtain:

$$\frac{\partial e}{\partial t} + \text{div}\,(\tilde{\mathbf{S}} + \hat{\mathbf{S}}) = \mathbf{b}\dot{\mathbf{u}}. \tag{3.17}$$

In the absence of external forces, i.e. $\mathbf{b} = 0$, it takes the form of a *continuity equation*:

$$\frac{\partial e}{\partial t} + \text{div}\,(\tilde{\mathbf{S}} + \hat{\mathbf{S}}) = 0. \tag{3.18}$$

Since the increase of energy density has got to be equal to the inflow of energy per unit volume, $\tilde{\mathbf{S}} + \hat{\mathbf{S}}$ can be interpreted as the *energy flow,* and the two newly defined properties then become the energy flow vectors of solenoidal and irrotational energy respectively. I choose to define the energy flow as two different vectors because I am going to treat them separately in later discussions.

3.9   Sinks and sources

A *sink* is defined as a point, or singularity, into which spatial mass is
flowing, and accordingly, a *source* is a point with an outflow of mass.
Sinks and sources can be used as mathematical tools to describing
bodies moving without turbulence through a gas or liquid. As an
example, the superposition of the flow around a dipole of one source
and one equally strong sink and a steady flow in the direction of
the dipole axis, will give a perfect picture of the flow around a solid
ball in the liquid.

In Section 3.6 I found that by small deformations, spatial deformations can alternatively be treated as being solenoidal (no divergence) or irrotational (no shear). In the last representation space
behaves much like a fluid and can be treated as such. Spatial mass
can of course not be created at the source and disappear at the sink,
so for a sink source duple to work, there need to be some kind of
a loophole bringing spatial mass back from the sink to the source.
That is the role of virtual photons in the space between the two
items as will be discussed in Section 3.9 (see also Sec. 5.7 and 7.3).
They travel back and forth in the space between the two items and
bring chunks of spatial mass or holes with them in order to keep
the sources and sinks going. For now it is enough to assume that
such free moving objects exist, and that they only can be created
and disappear by pair production and annihilation — one sink for
one equally strong source and vice versa.

In this section I shall discuss how such hypothetical free moving
entities would perform in a solenoidal deformation field. I will take
sinks as positive entities and sources as negative sinks. If there are
more sinks than sources in an area, the *sink density* is positive,
and if there are more sources than sinks, then the sink density is
negative.

If we integrate the energy in the velocity field around a source
or a sink from a small radius, ϵ, to infinity, and let ϵ shrink towards
zero, then the field energy would approach infinity. As sinks and
sources so far are only hypothetical object to which we are free to
ascribe any characteristics of our choice, we could tentatively say

that their sizes are big enough to make the field energy around them finite.

The strength of a sink, Q_s, can be defined as the inflow of spatial mass through a closed surface around the sink

$$Q = -\rho_s \oint_V \dot{\mathbf{u}} \cdot \mathbf{n} \, df, \qquad (3.19)$$

where $\mathbf{n}$ is an outwards pointing unit vector normal to df.

Now let sinks and sources with strength $Q_1, Q_2, \cdots, Q_n$ be sufficiently smoothly distributed in space. Then the *density of sinks*, ρ_q,[7] is given by

$$\begin{aligned}
\rho_q &= \lim_{V \to \epsilon} \frac{1}{V} \sum_{n=1}^{m} Q_n \\
&= -\rho_s \lim_{V \to \epsilon} \frac{1}{V} \oint_V \dot{\mathbf{u}} \cdot \mathbf{n} df \\
&= -\rho_s \mathrm{div}\, \dot{\mathbf{u}}, \\
\mathrm{div}\, \dot{\mathbf{u}} &= -\frac{\rho_q}{\rho_s}, \qquad (3.20)
\end{aligned}$$

where m is the number of sinks in a volume V of space, and ϵ is a small volume, but still great enough to contain many sinks.

There is a *hidden dependency* between the sink density ρ_q and and the volume force $\mathbf{b}$. By taking the divergence of the solenoidal N-C Equation, (3.10), we obtain[8]:

$$\begin{aligned}
-\mathrm{div}\,(\mathrm{curl}\,\mathrm{curl}\,\tilde{\mathbf{u}}) + \mathrm{div}\,\frac{\tilde{\mathbf{b}}}{\mu_s} &= \frac{1}{c_l^2} \mathrm{div}\,\ddot{\tilde{\mathbf{u}}} \\
\mathrm{div}\,\ddot{\tilde{\mathbf{u}}} &= \frac{c_l^2}{\mu_s} \mathrm{div}\,\tilde{\mathbf{b}} \\
\frac{\partial}{\partial t}(\mathrm{div}\,\dot{\mathbf{u}}) &= \frac{1}{\rho_s} \mathrm{div}\,\tilde{\mathbf{b}}, \qquad (3.21)
\end{aligned}$$

---

[7]Notice that ρ_q with the index q is not like the spatial mass density ρ_s but another property, the density of sinks.
[8]Notice that even if we have introduced 'pointlike' sinks and sources into the spatial continuum, the space itself may well be considered divergence-free.

and by taking the partial derivative with respect on time of (3.20), we get:

$$\frac{\partial}{\partial t}(\text{div } \dot{\mathbf{u}}) \; = \; -\frac{\dot{\rho}_q}{\rho_s} \, . \tag{3.22}$$

By evaluating the combination, (3.21) – (3.22), we obtain:

$$\dot{\rho}_q + \text{div } \mathbf{b} = 0 \, . \tag{3.23}$$

Since sinks and sources by definition only can be created or disappear in pairs, the only way the sink density can change in a volume is by out- or inflow, hence the vector $\mathbf{b}$ can be interpreted as a flow of sinks or sources, and conversely, a flow of sinks or sources will create a body force (a push or a drag) in the spatial continuum.

3.10   Solenoidal energy and momentum

In this section we set $\hat{u} = 0$, so $u = \tilde{u}$ and the energy flow vector (3.10) reduces to:

$$\tilde{\mathbf{S}} = -\mu_s \dot{\mathbf{u}} \times \text{curl } \mathbf{u} \, . \tag{3.24}$$

The time derivative of the energy flow is:

$$\frac{\partial \tilde{\mathbf{S}}}{\partial t} = -\mu_s \dot{\mathbf{u}} \times \text{curl } \dot{\mathbf{u}} - \mu_s \ddot{\mathbf{u}} \times \text{curl } \mathbf{u} \, .$$

I treat the second term with the mathematical identity:

$$\mathbf{A} \times \text{curl } \mathbf{A} \; = \; \frac{1}{2}\text{grad}\,(\mathbf{A} \cdot \mathbf{A}) - (\mathbf{A} \cdot \nabla)\mathbf{A} \, ,$$

and from (3.10) I plug in:

$$\ddot{\mathbf{u}} = -c_l^2 \text{curl curl } \mathbf{u} + \frac{\mathbf{b}}{\rho_s} \, .$$

We obtain

$$\dot{\mathbf{S}} = -\mu_s \left[\frac{1}{2} \text{grad} \, (\dot{\mathbf{u}} \cdot \dot{\mathbf{u}}) - (\dot{\mathbf{u}} \cdot \nabla) \dot{\mathbf{u}} \right]$$

$$-\mu_s \left[-c_l^2 \text{curl curl} \, \mathbf{u} + \frac{\mathbf{b}}{\rho_s} \right] \times \text{curl} \, \mathbf{u},$$

$$\frac{\dot{\mathbf{S}}}{c_l^2} = \frac{\mu_s}{c_l^2} (\dot{\mathbf{u}} \cdot \nabla) \dot{\mathbf{u}} - \frac{\mu_s}{2c_l^2} \text{grad} \, (\dot{\mathbf{u}} \cdot \dot{\mathbf{u}})$$

$$-\mu_s \text{curl} \, \mathbf{u} \times \text{curl curl} \, \mathbf{u} + \text{curl} \, \mathbf{u} \times \mathbf{b}.$$

By the same mathematical identity as above and $c_l^2 = \mu_s / \rho_s$, we get:

$$\frac{\dot{\mathbf{S}}}{c_l^2} = \rho_s (\dot{\mathbf{u}} \cdot \nabla) \dot{\mathbf{u}} - \text{grad} \left(\frac{\rho_s}{2} \dot{\mathbf{u}}^2 \right)$$

$$-\text{grad} \left[\frac{\mu_s}{2} (\text{curl} \, \mathbf{u})^2 \right] + \mu_s (\text{curl} \, \mathbf{u} \cdot \nabla) \text{curl} \, \mathbf{u} + \text{curl} \, \mathbf{u} \times \mathbf{b}.$$

According to (3.14) the *grad*-terms are the energy density, e, of the packet, so:

$$\frac{\dot{\mathbf{S}}}{c_l^2} + \text{grad} \, e - \mu_s (\text{curl} \, \mathbf{u} \cdot \nabla) \text{curl} \, \mathbf{u}$$

$$-\rho_s (\dot{\mathbf{u}} \cdot \nabla) \dot{\mathbf{u}} = \text{curl} \, \mathbf{u} \times \mathbf{b}.$$

This equation can be written as (quasi) components of the vectors $\mathbf{S}$, $\mathbf{b}$, $\dot{\mathbf{u}}$, curl$\mathbf{u}$, and curl $\mathbf{u} \times \mathbf{b}$. Remember also that we are dealing with a solenoidal field where $\mathbf{u}$ can be expressed as the curl of some vector field $\mathbf{A}$, see (3.8), hence div $\mathbf{u} = 0$ because the divergence of a curl is always zero. We obtain:

$$\frac{\dot{S}_a}{c^2} + e_{,a} - \rho_s \dot{u}_b \dot{u}_{a,b} - \mu_s (\text{curl} \, \mathbf{u})_b (\text{curl} \, \mathbf{u})_{a,b} = (\text{curl} \, \mathbf{u} \times \mathbf{b})_a \,,$$

$$\frac{\dot{S}_a}{c^2} + e_{,a} - \rho_s (\dot{u}_b \dot{u}_a)_{,b} + \rho_s \dot{u}_{b,b} \dot{u}_a - \mu_s [(\text{curl} \, \mathbf{u})_b (\text{curl} \, \mathbf{u})_a]_{,b}$$
$$+\mu_s [(\text{curl} \, \mathbf{u})_{b,b} (\text{curl} \, \mathbf{u})_a] = (\text{curl} \, \mathbf{u} \times \mathbf{b})_a \,.$$

The terms, $\rho_s \dot{u}_{b,b} \dot{u}_a$, and, $\mu_s [(\text{curl} \, \mathbf{u})_{b,b} (\text{curl} \, \mathbf{u})_a]$, vanishes because the divergence of any curl is always zero. We also have that $\delta_{abe,b} =$

$e_{,a}$ where δ_{ab} is the Kronecker delta[9]. We obtain:

$$\frac{\dot{\tilde{S}}_a}{c^2} - [-\delta_{ab}e + \rho_s \dot{u}_b \dot{u}_a + \mu_s(\text{curl}\,\mathbf{u})_b(\text{curl}\,\mathbf{u})_a]_{,b} = (\text{curl}\,\mathbf{u} \times \mathbf{b})_a$$

The expression inside the square bracket is a second order tensor which I can set like a new property

$$\tilde{\sigma}_{ab} = -\delta_{ab}e + \rho_s \dot{u}_a \dot{u}_b + \mu_s(\text{curl}\,\mathbf{u})_a(\text{curl}\,\mathbf{u})_b, \qquad (3.25)$$

and we obtain

$$\frac{\dot{\tilde{S}}_a}{c^2} - \tilde{\sigma}_{ab,b} = (\text{curl}\,\mathbf{u} \times \mathbf{b})_a$$

Together with Equation (3.18) the equation above can be assembled to a group of equations:

$$-\frac{\partial e}{\partial t} - \frac{\partial \tilde{S}_x}{\partial x} - \frac{\partial \tilde{S}_y}{\partial y} - \frac{\partial \tilde{S}_z}{\partial z} = -\mathbf{b}\dot{\mathbf{u}},$$

$$\frac{\partial \tilde{S}_x}{c^2 \partial t} - \frac{\partial \tilde{\sigma}_{xx}}{\partial x} - \frac{\partial \tilde{\sigma}_{xy}}{\partial y} - \frac{\partial \tilde{\sigma}_{xz}}{\partial z} = (\text{curl}\,\mathbf{u} \times \mathbf{b})_x,$$

$$\frac{\partial \tilde{S}_y}{c^2 \partial t} - \frac{\partial \tilde{\sigma}_{yx}}{\partial x} - \frac{\partial \tilde{\sigma}_{yy}}{\partial y} - \frac{\partial \tilde{\sigma}_{yz}}{\partial z} = (\text{curl}\,\mathbf{u} \times \mathbf{b})_y,$$

$$\frac{\partial \tilde{S}_z}{c^2 \partial t} - \frac{\partial \tilde{\sigma}_{zx}}{\partial x} - \frac{\partial \tilde{\sigma}_{zy}}{\partial y} - \frac{\partial \tilde{\sigma}_{zz}}{\partial z} = (\text{curl}\,\mathbf{u} \times \mathbf{b})_z.$$

Now we define a couple of new mixed four tensors:

$$\tilde{T}_\alpha{}^\beta = \begin{bmatrix} -e & -\tilde{S}_x/c & -\tilde{S}_y c & -\tilde{S}_z/c \\ \tilde{S}_x/c & -\tilde{\sigma}_{xx} & -\tilde{\sigma}_{xy} & -\tilde{\sigma}_{xz} \\ \tilde{S}_y/c & -\tilde{\sigma}_{yx} & -\tilde{\sigma}_{yy} & -\tilde{\sigma}_{yz} \\ \tilde{S}_z/c & -\tilde{\sigma}_{zx} & -\tilde{\sigma}_{zy} & -\tilde{\sigma}_{zz} \end{bmatrix}, \qquad (3.26)$$

---

[9]We shall see that in the electromagnetic interpretation of these equations the negative of div $\dot{\mathbf{u}}$ represents the density of sinks, i.e. negative electrical charges, and cannot be left out.

$$
\tilde{D}_\alpha{}^\beta = \begin{bmatrix}
0 & -\dfrac{\dot{u}_x}{c} & -\dfrac{\dot{u}_y}{c} & -\dfrac{\dot{u}_z}{c} \\
\dfrac{\dot{u}_x}{c8} & 0 & -(\operatorname{curl}\mathbf{u})_z & (\operatorname{curl}\mathbf{u})_y \\
\dfrac{\dot{u}_y}{c} & (\operatorname{curl}\mathbf{u})_z & 0 & -(\operatorname{curl}\mathbf{u})_x \\
\dfrac{\dot{u}_z}{c} & -(\operatorname{curl}\mathbf{u})_y & (\operatorname{curl}\mathbf{u})_x & 0
\end{bmatrix},
$$

and the hypothetical external force:

$$
b^\alpha = \begin{bmatrix} 0 \\ b_x \\ b_y \\ b_z \end{bmatrix},
$$

With these properties, the group of equations above can compactly be written as:

$$
\tilde{T}_\mu{}^{\alpha,\mu} = \tilde{D}_\nu{}^\alpha b^\nu.
$$

Notice that the tensors are represented as mixed tensor. Therefore we at this stage don't need to bring any form of metrics into the equations.

3.11   Irrotational energy and momentum

In this section I set $\tilde{u} = 0$, so $u = \hat{u}$ and the energy flow vector becomes::

$$
\hat{\mathbf{S}} = -(\lambda_s + 2\mu_s)\dot{\hat{\mathbf{u}}} \cdot \operatorname{div}\mathbf{u},
$$

$$
\frac{1}{c_g{}^2}\hat{\mathbf{S}} + \rho_s\dot{\mathbf{u}} \cdot \operatorname{div}\mathbf{u} = 0.
$$

The time derivative of the energy flow becomes:

$$
\frac{1}{c_g{}^2}\frac{\partial\hat{\mathbf{S}}}{\partial t} + \rho_s\ddot{\mathbf{u}} \cdot \operatorname{div}\mathbf{u} + \rho_s\dot{\mathbf{u}} \cdot \operatorname{div}\dot{\mathbf{u}} = 0. \tag{3.27}
$$

By the identity:

$$
\operatorname{grad}(\mathbf{A} \cdot \mathbf{A}) = 2\left[\mathbf{A} \times \operatorname{curl}\mathbf{A} + (\mathbf{A} \cdot \nabla)\mathbf{A}\right],
$$

we generally can deduce the relation:

$$
\begin{aligned}
(\dot{u}_a \dot{u}_b)_{,b} &= \dot{u}_a \dot{u}_{b,b} + \dot{u}_b \dot{u}_{a,b} \\
&= \dot{u}_a \operatorname{div} \dot{u} + [(\dot{u}\nabla)\dot{u}]_a, \\
&= \dot{u}_a \operatorname{div} \dot{u} + [\tfrac{1}{2}\operatorname{grad}(\dot{u} \cdot \dot{u}) - \dot{u} \times \operatorname{curl} \dot{u}]_a, \\
\dot{u}_a \cdot \operatorname{div} \dot{u} &= (\dot{u}_a \dot{u}_b)_{,b} - \frac{1}{2}(\dot{u} \cdot \dot{u})_{,a} + (\dot{u} \times \operatorname{curl} \dot{u})_a.
\end{aligned}
$$

The last term above disappears because $\operatorname{curl} \dot{u} = 0$, so:

$$
\dot{u}_a \cdot \operatorname{div} \dot{u} = (\dot{u}_a \dot{u}_b)_{,b} - \tfrac{1}{2}(\dot{u} \cdot \dot{u})_{,a}.
$$

From the irrotational part of the N-C equation, multiplied by $\operatorname{div} \mathbf{u}$, we obtain:

$$
\ddot{u} = c_g{}^2 \operatorname{grad} \operatorname{div} \mathbf{u} + \frac{\mathbf{b}}{\rho_s} \quad \Big| \cdot \operatorname{div} \mathbf{u},
$$

$$
\begin{aligned}
(\ddot{u} \operatorname{div} \mathbf{u})_a &= c_g{}^2 (\operatorname{div} \mathbf{u})(\operatorname{grad} \operatorname{div} \mathbf{u})_{,a} + \frac{1}{\rho_s} b_a \operatorname{div} \mathbf{u}, \\
&= \frac{1}{2} c_g{}^2 \operatorname{grad}(\operatorname{div}{}^2 \mathbf{u})_{,a} + \frac{1}{\rho_s} b_a \operatorname{div} \mathbf{u}.
\end{aligned}
$$

With these properties inserted, (3.27) becomes:

$$
\frac{1}{c_g{}^2} \dot{\hat{S}}_a + (\lambda_s + 2u_s)(\operatorname{grad} \operatorname{div} \mathbf{u})_a \cdot \operatorname{div} \mathbf{u} + \rho_s(\dot{u}_a \dot{u}_b)_{,b} - \frac{\rho_s}{2}(\dot{u} \cdot \dot{u})_{,a}
$$

$$
= b_a \cdot \operatorname{div} \mathbf{u},
$$

$$
\frac{1}{c_g{}^2} \dot{\hat{S}}_a + \frac{(\lambda_s + 2u_s)}{2}(\operatorname{div} \mathbf{u} \cdot \operatorname{div} \mathbf{u})_{,a} + \rho_s(\dot{u}_a \dot{u}_b)_{,b} - \frac{\rho_s}{2}(\dot{u} \cdot \dot{u})_{,a}
$$

$$
= b_a \cdot \operatorname{div} \mathbf{u}.
$$

The right side of the equation, involving influence from the outside world, is brought along mainly as a place holder, and may be set to zero if there is no such influence.

The total energy of the field is like the sum of kinetic and potential energy, $\hat{e} = \hat{e}_k + \hat{e}_p$, where the two components are given by:

$$\hat{e}_k = \frac{\rho_s}{2}(\dot{\mathbf{u}} \cdot \dot{\mathbf{u}}),$$

$$\hat{e}_p = \frac{(\lambda_s + 2u_s)}{2}(\operatorname{div}\mathbf{u} \cdot \operatorname{div}\mathbf{u}),$$

respectively. Then the equation above takes the form:

$$\frac{1}{c_g^2}\dot{\hat{S}}_a + \rho_s(\dot{u}_a\dot{u}_b)_{,b} - (\hat{e}_k)_{,a} + (\hat{e}_p)_{,a} = b_a \cdot \operatorname{div}\mathbf{u},$$

$$\frac{\partial\hat{S}_a}{c_g^2\partial t} + \rho_s(\dot{u}_a\dot{u}_b)_{,b} - \delta_{ab}(\hat{e}_k - \hat{e}_p)_{,b} = b_a \cdot \operatorname{div}\mathbf{u},$$

which by defining a new tensor:

$$\hat{\sigma}_{ab} = \rho_s(\dot{u}_a\dot{u}_b) + \delta_{ab}(\hat{e}_p - \hat{e}_k), \tag{3.28}$$

reduces to:

$$\frac{\partial\hat{S}_a}{c_g^2\partial t} + \hat{\sigma}_{ab,b} = b_a \cdot \operatorname{div}\mathbf{u},$$

Equation 3.18,

$$\frac{\partial\hat{e}}{\partial t} + \operatorname{div}\hat{\mathbf{S}} = \mathbf{b} \cdot \dot{\mathbf{u}},$$

and the equation above can be put together to a group of equations:

$$-\frac{\partial\hat{e}}{c_g\partial t} - \frac{\partial}{\partial x}(\frac{\hat{S}_x}{c_g}) - \frac{\partial}{\partial y}(\frac{\hat{S}_y}{c_g}) - \frac{\partial}{\partial z}(\frac{\hat{S}_z}{c_g}) = -\frac{1}{c_g}\dot{\mathbf{u}} \cdot \hat{\mathbf{b}},$$

$$\frac{\partial}{c_g\partial t}(\frac{\hat{S}_x}{c_g}) + \frac{\partial\hat{\sigma}_{xx}}{\partial x} + \frac{\partial\hat{\sigma}_{xy}}{\partial y} + \frac{\partial\hat{\sigma}_{xz}}{\partial z} = b_x\operatorname{div}\mathbf{u},$$

$$\frac{\partial}{c_g\partial t}(\frac{\hat{S}_y}{c_g}) + \frac{\partial\hat{\sigma}_{yx}}{\partial x} + \frac{\partial\hat{\sigma}_{yy}}{\partial y} + \frac{\partial\hat{\sigma}_{yz}}{\partial z} = b_y\operatorname{div}\mathbf{u},$$

$$\frac{\partial}{c_g\partial t}(\frac{\hat{S}_x}{c_g}) + \frac{\partial\hat{\sigma}_{zx}}{\partial x} + \frac{\partial\hat{\sigma}_{zy}}{\partial y} + \frac{\partial\hat{\sigma}_{zz}}{\partial z} = b_z\operatorname{div}\mathbf{u}.$$

Again we define a couple of new mixed four tensors:

$$
\hat{T}_\alpha{}^\beta =
\begin{bmatrix}
-\hat{e} & -\dfrac{\hat{S}_x}{c_g} & -\dfrac{\hat{S}_y}{c_g} & -\dfrac{S_z}{c_g} \\[2mm]
\dfrac{\hat{S}_x}{c_g} & \hat{\sigma}_{xx} & \hat{\sigma}_{xy} & \hat{\sigma}_{xz} \\[2mm]
\dfrac{\hat{S}_y}{c_g} & \hat{\sigma}_{yx} & \hat{\sigma}_{yy} & \hat{\sigma}_{yz} \\[2mm]
\dfrac{\hat{S}_z}{c_g} & \hat{\sigma}_{zx} & \hat{\sigma}_{zy} & \hat{\sigma}_{zz}
\end{bmatrix},
\tag{3.29}
$$

$$
\hat{D}_\alpha{}^\beta =
\begin{bmatrix}
0 & -\dfrac{\dot{u}_x}{c_g} & -\dfrac{\dot{u}_y}{c_g} & -\dfrac{\dot{u}_z}{c_g} \\[2mm]
\dfrac{\dot{u}_x}{c_g} & \operatorname{div}\mathbf{u} & 0 & 0 \\[2mm]
\dfrac{\dot{u}_y}{c_g} & 0 & \operatorname{div}\mathbf{u} & 0 \\[2mm]
\dfrac{\dot{u}_z}{c_g} & 0 & 0 & \operatorname{div}\mathbf{u}
\end{bmatrix},
$$

Then the group of equations above can be written compactly as:

$$
\partial^\mu \hat{T}_\mu{}^\alpha = b^\nu \hat{D}_\nu{}^\alpha.
$$

Like with the solenoidal field, the tensor $\hat{\mathbf{T}}$ can be interpreted as the stress energy tensor for the irrotational field (see 3.26). The two stress energy tensors can be added together to give the resulting stress energy tensor:

$$
T_\alpha{}^\beta = \tilde{T}_\alpha{}^\beta + \hat{T}_\alpha{}^\beta.
\tag{3.30}
$$

This relation will reveal a possible connection between the solenoidal and the irrotational field. Finally, note again that we only are dealing with mixed tensors and do not yet need a metric.

Chapter 4

Electrodynamics

Already in the Nineteenth century *William Thomson (Lord Kelvin) (1824–1907)* and others pointed out the resemblance between elastodynamic and electrodynamic equations [15, page 279-280]. In this chapter I shall follow up some of these thoughts in order to see exactly how far the resemblance stretches. First I'll compare the two disciplines term by term and then go a bit further to see if the comparison holds also for more elaborate electrodynamic properties. In the last part of the chapter, Sec. 4.6, I'll deduce the most general form of Maxwell's equations directly from the Navier-Cauchy equation into a form that is invariant under Lorenz transformations.

4.1 Maxwell's equations

For the rest of this chapter I'll only consider a divergence-free spatial continuum. Hence $\mathbf{u} = \tilde{\mathbf{u}}$ and div $\mathbf{u} = 0$.

First I define some new properties:

$$\mu_0 = \frac{1}{\mu_s}, \tag{4.1}$$

$$\varepsilon_0 = \rho_s, \tag{4.2}$$

$$\mathbf{E} = -\frac{\partial \tilde{\mathbf{u}}}{\partial t}, \tag{4.3}$$

$$\mathbf{B} = \mathrm{curl}\,\tilde{\mathbf{u}}, \tag{4.4}$$

$$\mathbf{j} = \mathbf{b}, \tag{4.5}$$

$$c^2 = \frac{\mu_s}{\rho_s} = \frac{1}{\varepsilon_0 \mu_0}. \tag{4.6}$$

By applying the curl operator on $\mathbf{E}$,

$$\mathrm{curl}\,\mathbf{E} = \mathrm{curl}\left(-\frac{\partial \tilde{\mathbf{u}}}{\partial t}\right)$$

$$= -\frac{\partial}{\partial t}\mathrm{curl}\,\tilde{\mathbf{u}},$$

we obtain:

$$\mathrm{curl}\,\mathbf{E} + \dot{\mathbf{B}} = 0 \tag{4.7}$$

We also have the identity that the divergence of a curl is zero, so:

$$\mathrm{div}\,\mathbf{B} = 0 \tag{4.8}$$

With the newly defined terms inserted, the Navier-Cauchy equation for solenoidal deformations, (3.9), takes the form:

$$\mathrm{curl}\,\mathbf{B} - \frac{1}{c^2}\dot{\mathbf{E}} = \mu_0 \mathbf{j} \tag{4.9}$$

Finally I shall adopt the idea that electric charges and charge density behave like sinks and sources in the spatial continuum (see Sec 5.7 and 5.8). Then an electric charge takes the form of a sink, and according to (3.19), the strength of a charge in the volume element V, becomes:

$$Q = \epsilon_0 \oint_V \mathbf{E} \cdot \mathbf{n} df.$$

The charge density can be associated with the density of sinks, and by (3.20), the corresponding electrodynamic equation becomes:

$$\dot{\rho}_q + \operatorname{div} \mathbf{j} = 0 \tag{4.10}$$

Here ρ_q is to be interpreted as electric charge density, and $\mathbf{j}$ as the current.

The four equations (4.7) ... (4.10) above are the *Maxwell equations* in their most common representation, so if elementary electric charges can be found to act like sinks and sources in a spatial continuum, we would have at our hands a mechanical model of electrodynamics. A point-like sink or source, however, would be surrounded by a velocity field energy that approaches infinity, but here it is enough to make the statement that charges *perform* like sinks and sources in some small volume of space. Note that so far the model is only valid in a Cartesian coordinate system at rest in space.

4.2 The stress-energy tensor

According to (3.14) and the newly defined properties the elastodynamic field energy in a divergence-free field is

$$e = \tfrac{\varepsilon_0}{2}\mathbf{E}^2 + \tfrac{1}{2\mu_0}\mathbf{B}^2. \tag{4.11}$$

Pointing's vector in a solenoid deformation field, is a direct consequence of the energy flow vector, (3.24):

$$\mathbf{S} = \frac{1}{\mu_0}\mathbf{E} \times \mathbf{B}, \tag{4.12}$$

and the corresponding form of Pointing's theorem follows directly:

$$\frac{\partial e}{\partial t} + \operatorname{div} \mathbf{S} = -\mathbf{j} \cdot \mathbf{E}. \tag{4.13}$$

Equation (3.25) gives us the expression for the Mawell's stress tensor:

$$\sigma_{ij} = -\delta_{ij}e + \epsilon_0 E_i E_j + \frac{1}{\mu_s}B_i B_j, \tag{4.14}$$

and the time derivative of **S** becomes:

$$\frac{\dot{S}_i}{c^2} - \sigma_{ij,j} = -\epsilon_0 E_{j,j} E_i + \epsilon_{ijk} B_j b_k \qquad (4.15)$$

As I argued with the elastic equations, I set both ρ_q and **j** to zero in order to find the meaning of σ:

$$\frac{1}{c^2} d\mathbf{S} = \operatorname{div} \sigma dt,$$

and find that the momentum of the energy flow is:

$$\mathbf{p} = \frac{\mathbf{S}}{c^2}. \qquad (4.16)$$

S is the energy that per unit time passes through a square unit of space. It relates to an energy density, e, moving with the speed of **v** by the relation: $\mathbf{S} = e\mathbf{v}$. Hence if a quantum of energy, E, is moving along with the speed of light, it will have a momentum given by:

$$\mathbf{p} = \frac{E}{c}\mathbf{n}, \qquad (4.17)$$

where **n** is a unit vector in the direction of movement.

Now we write out (4.13) and (4.15) in component form and obtain the set of equations (the zeroes are inserted for clarity):

$$\frac{\partial e}{c\partial t} + \frac{\partial S_x}{c\partial x} + \frac{\partial S_y}{c\partial y} + \frac{\partial S_z}{c\partial z} = \quad 0 - \frac{E_x j_x}{c} - \frac{E_y j_y}{c} - \frac{E_z j_z}{c},$$

$$\frac{\partial S_x}{c^2 \partial t} - \frac{\partial \sigma_{xx}}{\partial x} - \frac{\partial \sigma_{xy}}{\partial y} - \frac{\partial \sigma_{xz}}{\partial z} = \quad -E_x \rho_q + 0 + B_z j_y - B_y j_z,$$

$$\frac{\partial S_y}{c^2 \partial t} - \frac{\partial \sigma_{yx}}{\partial x} - \frac{\partial \sigma_{yy}}{\partial y} - \frac{\partial \sigma_{yz}}{\partial z} = \quad -E_y \rho_q - B_z j_x + 0 + B_x j_z,$$

$$\frac{\partial S_z}{c^2 \partial t} - \frac{\partial \sigma_{zx}}{\partial x} - \frac{\partial \sigma_{zy}}{\partial y} - \frac{\partial \sigma_{zz}}{\partial z} = \quad -E_z \rho_q + B_y j_x - B_x j_y + 0.$$

These four equations can be drawn together to one matrix equation:

$$
\begin{bmatrix} \frac{\partial}{c\partial t} & \frac{\partial}{\partial x} & \frac{\partial}{\partial y} & \frac{\partial}{\partial z} \end{bmatrix} \cdot
\begin{bmatrix}
e & S_x/c & S_y/c & S_z/c \\
S_x/c & -\sigma_{xx} & -\sigma_{yx} & -\sigma_{zx} \\
S_y/c & -\sigma_{xy} & -\sigma_{yy} & -\sigma_{zy} \\
S_z/c & -\sigma_{xz} & -\sigma_{yz} & -\sigma_{zz}
\end{bmatrix}
$$

$$
=
\begin{bmatrix}
0 & -E_x/c & -E_y/c & -E_z/c \\
E_x/c & 0 & B_z & -B_y \\
E_y/c & -B_z & 0 & B_x \\
E_z/c & B_y & -B_x & 0
\end{bmatrix}
\cdot
\begin{bmatrix}
-c\rho_q \\
j_x \\
j_y \\
j_z
\end{bmatrix}
\qquad (4.18)
$$

4.3   The flat Minkowski space

The spatial continuum is supposed to fill all space. It is deformable and the deformation can be described by the displacement vector **u**, which will vary throughout space, and hence make up a deformation field. If we consider the displacement vector at one particular point in space, it will have a length and a direction, but in order to measure it properly, we will need some grid to measure it against. So we define three orthogonal unit vectors, e_1, e_2, e_3. Normal to each axis we place an invisible plane at unit distances from each other, and number them such that plan number zero goes through the origin. We can compare the set-up with a hypothetical, sufficiently fine-meshed cubical grid with edges of unit lengths, almost like an old time egg-crate with one cell for each egg. We can now measure the length of a vector by counting the number of walls it penetrates in the three directions. These are the coordinates of the vector. Next we put in a coarser meshed grid and measure the coordinates in that grid. The vector itself does not change during this conversion, but since the net is coarser, the vector spans over fewer surfaces, so while the edges of meshes become larger the coordinate numbers become smaller. Therefore such vectors are called contravariant vectors, and we mark them by putting the coordinate indices in the upstairs position.

Another property we could study is the density of the spatial continuum itself. Perhaps it is compressed at some places and expanded elsewhere. That would make up a scalar field which vary as a function of position. We could construct a number for the change of density over a certain number of egg-cells. Measurements in the coarser grid, however, would give us a higher number because the measured distance is greater. This number varies along with the grid, and hence it is a covariant property. Mathematically such conditions can be described by taking the gradient of the density, which produces a vector. Since this vector is covariant, we put the indices in the downstairs position.

In Euclidean space the distinction between co- and contravariant vectors does not make any difference, so in such a space all the indices are placed in a downstairs position and denoted by italic indices. A vector is defined as:

$$\mathbf{v} = v_a \mathbf{e}_a,$$

the *Euclidean metric* by

$$\delta_a^b = \mathbf{e}_a \mathbf{e}^b = \begin{bmatrix} 1 & 0 & 0 \\ 0 & 1 & 0 \\ 0 & 0 & 1 \end{bmatrix},$$

and the dot product is given by

$$\mathbf{v} \cdot \mathbf{w} = \delta_a^b v_a w_b.$$

In Minkowski space space the time component is added to the position coordinates:

$$x^\alpha = (x^0, x^1, x^2, x^3) = (ct, x, y, z).$$

A contravariant vector is given by

$$\mathbf{v} = v^0 \mathbf{e}_0 + v^1 \mathbf{e}_1 + v^2 \mathbf{e}_2 + v^3 \mathbf{e}_3 = v^\alpha \mathbf{e}_\alpha,$$

and accordingly a covariant vector, a covector, by

$$\mathbf{w} = w_0 \mathbf{e}^0 + w_1 \mathbf{e}^1 + w_2 \mathbf{e}^2 + w_3 \mathbf{e}^3 = w_\alpha \mathbf{e}^\alpha.$$

The *Minkowski metric* is

$$\eta_{\alpha\beta} = \mathbf{e}_\alpha \mathbf{e}_\beta = \begin{bmatrix} -1 & 0 & 0 & 0 \\ 0 & 1 & 0 & 0 \\ 0 & 0 & 1 & 0 \\ 0 & 0 & 0 & 1 \end{bmatrix} = \eta^{\alpha\beta},$$

and by applying the Einstein summation convention, the dot product between two vectors becomes:

$$\begin{aligned} \mathbf{a} \cdot \mathbf{b} &= \eta_{\alpha\beta} a^\alpha b^\beta = a^\alpha b_\alpha \\ &= -a^0 b_0 + a^1 b_1 + a^2 b_2 + a^3 b_3. \end{aligned}$$

We notice that the metric can be applied to lower and raise the indices.

The *del operator* is

$$\begin{aligned} \partial_\alpha &= \mathbf{e}^\alpha \frac{\partial}{\partial x^\alpha} \\ &= \mathbf{e}^0 \frac{\partial}{\partial x^0} + \mathbf{e}^1 \frac{\partial}{\partial x^1} + \mathbf{e}^2 \frac{\partial}{\partial x^2} + \mathbf{e}^3 \frac{\partial}{\partial x^3}. \end{aligned}$$

and

$$\begin{aligned} \partial^\alpha &= \eta^{\alpha\beta} \partial_\beta \\ &= -\mathbf{e}^0 \frac{\partial}{\partial x^0} + \mathbf{e}^1 \frac{\partial}{\partial x^1} + \mathbf{e}^2 \frac{\partial}{\partial x^2} + \mathbf{e}^3 \frac{\partial}{\partial x^3}. \end{aligned}$$

It can alternatively be written with the comma notation, or the ∇ sign, for example:

$$\partial_\alpha =_{,\alpha} = \nabla_\alpha.$$

Finally the *d'Alembert operator* is given by

$$\begin{aligned} \partial^\alpha \partial_\alpha &= \eta^{\alpha\beta} \partial_\beta \partial_\alpha, \\ \Box &= -\frac{\partial^2}{\partial (ct)^2} + \frac{\partial^2}{\partial (x)^2} + \frac{\partial^2}{\partial (y)^2} + \frac{\partial^2}{\partial (z)^2} \end{aligned}$$

4.4   Elastodynamics in four-space

In line with the notation treated above, Equation 4.18 can be written as:

$$T^{\alpha\beta}{}_{,\beta} = F^{\alpha\beta} J_\beta, \tag{4.19}$$

where

$$T^{\alpha\beta} = \begin{bmatrix} e & S_x/c & S_y/c & S_z/c \\ S_x/c & -\sigma_{xx} & -\sigma_{xy} & -\sigma_{xz} \\ S_y/c & -\sigma_{yx} & -\sigma_{yy} & -\sigma_{yz} \\ S_z/c & -\sigma_{zx} & -\sigma_{zy} & -\sigma_{zz} \end{bmatrix}, \tag{4.20}$$

is the *Minkowski Stress-energy tensor*, and

$$F^{\alpha\beta} = \begin{bmatrix} 0 & -E_x/c & -E_y/c & -E_z/c \\ E_x/c & 0 & -B_z & +B_y \\ E_y/c & +B_z & 0 & -B_x \\ E_z/c & -B_y & +B_x & 0 \end{bmatrix}, \tag{4.21}$$

is the *Electromagnetic tensor*.

The covector:

$$J_\alpha = (-c\rho_q, j_x, j_y, j_z), \tag{4.22}$$

is the *4-current*.

In this notation (4.10) on page 59,

$$\dot{\rho}_q + \operatorname{div} \mathbf{j} = 0,$$

can be written as

$$\partial_\alpha J^\alpha = 0,$$

where the contravector **J** is given by:

$$J^\alpha = (c\rho_q, j_x, j_y, j_z). \tag{4.23}$$

By comparing (4.22) and (4.23) we notice that the Minkowski metric indeed has been used to raise and lower the indices:

$$J^\alpha = \eta^{\alpha\beta} J_\beta,$$
$$J_\alpha = \eta_{\alpha\beta} J^\beta.$$

We conclude that (4.19) can be written in frame independent vector form:

$$\nabla \cdot \mathbf{T} = \mathbf{F} \cdot \mathbf{J}.$$

where $\mathbf{T}$ is the *Electromagnetic stress energy tensor*, $\mathbf{F}$ is the *Electromagnetic tensor*, and $\mathbf{J}$ is the *Four current*.

4.5 Trace of the stress energy tensor

The trace of a square matrix is an invariant property. Hence the trace of the electromagnetic stress energy tensor is of special interest.

From (4.21) we obtain the trace of $\mathbf{T}$ (tr $\mathbf{T}$):

$$\operatorname{tr} \mathbf{T} = T^\alpha{}_\alpha = -e - \sigma_{xx} - \sigma_{yy} - \sigma_{zz},$$

and from Maxwell's stress tensor (4.14) we obtain:

$$\sigma_{xx} = -e + \epsilon_0 E_x^2 + \frac{1}{\mu_0} B_x^2,$$
$$\sigma_{yy} = -e + \epsilon_0 E_y^2 + \frac{1}{\mu_0} B_y^2,$$
$$\sigma_{zz} = -e + \epsilon_0 E_z^2 + \frac{1}{\mu_0} B_z^2.$$

By (4.11) we finally obtain:

$$
\begin{aligned}
\operatorname{tr} \mathbf{T} &= -e + 3e - \epsilon_0(E_x^2 + E_y^2 + E_z^2) \\
&\quad - \frac{1}{\mu_0}(B_x^2 + B_y^2 + B_z^2) \\
&= -e + 3e - 2e = 0.
\end{aligned}
$$

The trace of the stress energy tensor turns out to be vanishing.

4.6   The vector potential

In the previous sections I have shown that the electromagnetic properties can be expressed in exactly the same way as elastodynamic equations in an elastic continuum of infinite extension. The correspondence is exact in an Euclidean coordinate system at rest in the spatial continuum. There are some good reasons to believe that it also is valid in a Lorenz frame in uniform rectilinear motion through space. In this section, however, I shall develop the electromagnetic equations directly from the elastodynamic equations without going the way around the electric and magnetic field equations. While the basic electromagnetic properties $\mathbf{E}$ and $\mathbf{B}$ clearly not are invariant by transformations between a resting frame and a frame in motion, I shall here go directly to properties that are Lorenz invariant.

Basic properties

I take point of departure from the Navier-Cauchy equation for a divergence-free deformation field (3.10) with μ_s set to $1/\mu_0$, (4.2), and $c = c_l = \sqrt{\mu_s/\rho_s}$:

$$-\frac{1}{c^2}\ddot{\mathbf{u}} - \operatorname{curl}\operatorname{curl}\mathbf{u} = -\mu_0\,\mathbf{b}. \tag{4.24}$$

As discussed in Section 3.9 on page 48 we presume that there may be moveable sinks and sources in the spatial continuum with a sink density of ρ_q given by (3.20)

$$
\begin{aligned}
\rho_q &= -\rho_s \operatorname{div}\dot{\mathbf{u}}, \\
\operatorname{div}\dot{\mathbf{u}} &= -\rho_q/\rho_s, \\
\operatorname{div}\dot{\mathbf{u}} &= -\mu_0 c^2 \rho_q.
\end{aligned} \tag{4.25}
$$

The dependency between the body force $\mathbf{b}$ and the sink density can be found by taking the divergence of the N-C equation (4.24)[1]:

$$
\begin{aligned}
-\frac{1}{c^2}\operatorname{div}\dddot{\mathbf{u}} - \operatorname{div}(\operatorname{curl}\operatorname{curl}\mathbf{u}) &= -\mu_0\operatorname{div}\mathbf{b}, \\
-\operatorname{div}\dddot{\mathbf{u}} &= -\mu_0 c^2 \operatorname{div}\mathbf{b},
\end{aligned}
$$

---

[1]The divergence of a curl is always zero.

and the time derivative of (4.25):

$$\text{div }\ddot{\mathbf{u}} = -\mu_0 c^2 \dot{\rho}_q.$$

By adding these two equations we obtain:

$$\text{div }\mathbf{b} + \dot{\rho}_q = 0.$$

Since sinks and sources by definition can only be created by pair production, or disappear by annihilation of one sink against one equally strong source, the change of sinks or sources inside a volume can only take place if there is an in- or outflow of sinks to the volume. Hence we can define a sink flow vector, $\mathbf{j}'$, given by the continuity equation:

$$\text{div }\mathbf{j}' + \frac{\partial \rho_q}{\partial t} = 0.$$

By comparing this equation with that above, we can conclude that in the absence of other body forces, the body force, $\mathbf{b}$, becomes equal to the sink flow vector and can be replaced with $\mathbf{j}'$ in the N-C equation. It becomes clear that a flow of sinks introduce a body force into the spatial continuum:

$$\mathbf{j}' = \mathbf{b}. \tag{4.26}$$

The displacement vector, $\mathbf{u}$, is defined in an Euclidean space, so the results so far are restricted to such frames. In an attempt to get at a more general description, I'll introduce a couple of new properties; a vector $\mathbf{A}'$ and temporarily a potential ψ, defined such that they satisfy the equation:

$$\mathbf{u} = \mathbf{A}' + \text{grad }\psi. \tag{4.27}$$

First I take the curl and the time derivative of (4.27):

$$\text{curl }\mathbf{u} = \text{curl }\mathbf{A}', \tag{4.28}$$
$$\dot{\mathbf{u}} = \dot{\mathbf{A}}' + \text{grad }\dot{\psi}. \tag{4.29}$$

Next I define a potential $\phi = \dot{\psi}$. Then I plug it into (4.29) and take the time derivative of the result. We obtain:

$$\ddot{\mathbf{u}} \;=\; \ddot{\mathbf{A}}' + \operatorname{grad} \dot{\phi}. \tag{4.30}$$

Finally I take the divergence of (4.29) and get:

$$\operatorname{div} \dot{\mathbf{u}} = \operatorname{div} \dot{\mathbf{A}}' + \nabla^2 \phi. \tag{4.31}$$

Compact electrodynamic equations

I am now in position to formally squeeze Maxwell's four equations into only one equation.

First I plug (4.28) and (4.30) into the N-C equation, (4.24), and obtain:

$$-\frac{1}{c^2}\ddot{\mathbf{A}}' - \operatorname{curl}\operatorname{curl}\mathbf{A}' - \frac{1}{c^2}\operatorname{grad}\dot{\phi} \;=\; -\mu_0\mathbf{j}'.$$

By the mathematical entity, $\operatorname{curl}\operatorname{curl}\mathbf{A} = \operatorname{grad}\operatorname{div}\mathbf{A} - \nabla^2\mathbf{A}$, we obtain:

$$-\frac{1}{c^2}\ddot{\mathbf{A}}' + \nabla^2\mathbf{A}' - \operatorname{grad}\operatorname{div}\mathbf{A}' - \frac{1}{c^2}\operatorname{grad}\dot{\phi} \;=\; -\mu_0\mathbf{j}',$$
$$-\frac{1}{c^2}\ddot{\mathbf{A}}' + \nabla^2\mathbf{A}' - \operatorname{grad}\left(\operatorname{div}\mathbf{A}' + \frac{1}{c^2}\dot{\phi}\right) \;=\; -\mu_0\mathbf{j}'.$$

Next I plug (4.31) into (4.25) and obtain:

$$\operatorname{div}\dot{\mathbf{u}} \;=\; -c^2\mu_0\rho_q,$$
$$\operatorname{div}\dot{\mathbf{A}}' + \nabla^2\phi \;=\; -c^2\mu_0\rho_q,$$
$$\operatorname{div}\dot{\mathbf{A}}' + \nabla^2\phi - \frac{1}{c^2}\ddot{\phi} + \frac{1}{c^2}\ddot{\phi} \;=\; -c^2\mu_0\rho_q,$$
$$-\frac{1}{c^2}\ddot{\phi} + \nabla^2\phi + \frac{\partial}{\partial t}\left(\operatorname{div}\mathbf{A}' + \frac{1}{c^2}\dot{\phi}\right) \;=\; -c^2\mu_0\rho_q.$$

I now fix the relation between $\mathbf{A}'$ and ϕ such that

$$\operatorname{div}\mathbf{A}' + \frac{1}{c^2}\dot{\phi} = 0. \tag{4.32}$$

Then the two equations above reduce to:

$$-\frac{1}{c^2}\left(\ddot{\phi}/c\right) + \nabla^2\left(\phi/c\right) = -\mu c\rho_q,$$

$$-\frac{1}{c^2}\ddot{\mathbf{A}}' + \nabla^2\mathbf{A}' = -\mu_0\mathbf{j}',$$

which can be written as the four equations:

$$-\frac{\partial^2(\phi/c)}{c^2\partial t^2} + \frac{\partial^2(\phi/c)}{\partial x^2} + \frac{\partial^2(\phi/c)}{\partial y^2} + \frac{\partial^2(\phi/c)}{\partial z^2} = -\mu_0(c\rho_q),$$

$$-\frac{\partial^2 A_x'}{c^2\partial t^2} + \frac{\partial^2 A_x'}{\partial x^2} + \frac{\partial^2 A_x'}{\partial y^2} + \frac{\partial^2 A_x'}{\partial z^2} = -\mu_0 j_x',$$

$$-\frac{\partial^2 A_y'}{c^2\partial t^2} + \frac{\partial^2 A_y'}{\partial x^2} + \frac{\partial^2 A_y'}{\partial y^2} + \frac{\partial^2 A_y'}{\partial z^2} = -\mu_0 j_y',$$

$$-\frac{\partial^2 A_z'}{c^2\partial t^2} + \frac{\partial^2 A_z'}{\partial x^2} + \frac{\partial^2 A_z'}{\partial y^2} + \frac{\partial^2 A_z'}{\partial z^2} = -\mu_0 j_z'.$$

With

$$A^\alpha = (\phi/c, A_x', A_y', A_z'),$$
$$J^\alpha = (c\rho_q, j_x', j_y', j_z'),$$

they become[2]:

$$-\frac{\partial^2 A^0}{c^2\partial t^2} + \nabla^2 A^0 = -\mu_0 J^0,$$

$$-\frac{\partial^2 A^a}{c^2\partial t^2} + \nabla^2 A^a = -\mu_0 J^a,$$

By d'Alembert's operator:

$$\Box = -\frac{\partial^2}{c^2\partial t^2} + \nabla^2,$$

the two equations can be compactly written as:

$$\Box\mathbf{A} = -\mu_0\mathbf{J}. \qquad (4.33)$$

---

[2]Roman letters run from 1-3.

which is Maxwell's equations in its most compact form. The relation (4.32) is the famous *Lorenz gauge*, which in four-space can be written:

$$\partial_\alpha A^\alpha = \partial^\alpha A_\alpha = 0.$$

The whole development above is a *gauge transformation* that allows us to rewrite the divergence-free elastodynamic equation of motion into a four-dimensional Minkowski space.

We define **E**, **B**, and **F**:

$$\begin{aligned}
\mathbf{E} &= -\dot{\mathbf{u}} = -\dot{\mathbf{A}} - \operatorname{grad}\phi, \\
\mathbf{B} &= \operatorname{curl}\mathbf{u} = \operatorname{curl}\mathbf{A}, \\
F^{\mu\nu} &= A^{\mu,\nu} - A^{\nu,\mu}.
\end{aligned}$$

The components of **F** can be found in this way:

$$\begin{aligned}
F^{00} &= -A^{0,0} - A^{0,0} = 0, \\
F^{01} &= A^{0,1} - A^{1,0} = \frac{\partial\phi}{c\partial x} + \frac{\partial A_x}{c\partial t} = -\frac{E_x}{c}, \\
F^{23} &= A^{2,3} - A^{3,2} = \frac{\partial A_y}{\partial z} - \frac{\partial A_z}{\partial y} = -(\operatorname{curl}\mathbf{A})_x = -B_x,
\end{aligned}$$

etc.

By computing all the components of **F**, we obtain:

$$F^{\alpha\beta} = \begin{bmatrix}
0 & -E_x/c & -E_y/c & -E_z/c \\
E_x/c & 0 & -B_z & +B_y \\
E_y/c & +B_z & 0 & -B_x \\
E_z/c & -B_y & +B_x & 0
\end{bmatrix},$$

which is seen to be the correct expression for the Electromagnetic tensor, **F**.

These equations are known to be invariant under Lorenz transformations, meaning that they will have the same form in any Lorenz frame moving with a constant speed in relation to each other.

Under the assumption that electric charges behave like free sinks and sources in an elastic, homogeneous, and isotropic continuum of infinite extension, this makes the mechanical model of electrodynamics complete.

Part III

Matter

Chapter 5

Matter as disturbance energy

And God said, "Let the waters under the sky be gathered together into one place, and let the dry land appear." (Genesis 1:9)

Say that we have established that light is some kind of wave movement through an elastic continuum, then what is matter? It simply cannot be any form of massive particles that force their way through the spatial continuum. The only viable answer to that question is that matter itself has got to be some kind of confined wave movement in the elastic continuum. Waves are energy, and so is mass. Einstein showed us that mass and energy are two sides of the same coin. So far so well, but mass consists of almost point-like particles, so we need to find wave patterns that meet also that kind of properties. That will be the topic of this chapter. Here I shall only outline the idea which will be considered to some more depths in the two next chapters.

5.1 A naïve model of matter

As an introduction to understanding the nature of matter, I shall first discuss a simple model based on confined wave energy. The simplest possible model of a material body one could think of, would be a weightless box with reflecting walls that keeps the waves within the box.

Confined wave energy will introduce a pressure into the spatial continuum given by 10.1:

$$p = \tfrac{1}{3}e,$$

and a body force given by 10.2:

$$\mathbf{b} = \operatorname{grad} p = \tfrac{1}{3}\operatorname{grad} e,$$

which can be inserted into the Navier-Cauchy equation. If the radiation is isotropic and the box is static, the equation takes the form:

$$(\lambda + 2\mu)\operatorname{grad}\operatorname{div}\mathbf{u} = \frac{1}{3}\operatorname{grad} e$$

$$\operatorname{div}\mathbf{u} = \frac{e}{3(\lambda + 2\mu)}$$

By the divergence theorem

$$\oint_V (\mathbf{A}\cdot\mathbf{n})df = \int_V \operatorname{div}\mathbf{A}dV,$$

we see that the displacement is independent of how densely the energy is distributed. Hence the total displacement from a volume where an amount of disturbance energy, E, is confined is given by

$$D = \frac{E}{3(\lambda+2\mu)}.$$

Figure 5.1: Wave energy in a box.

There are two kinds of waves that can occur in an elastic continuum, namely longitudinal compression waves and transversal shear waves (P- and S-waves). Both kinds of waves will carry energy and momentum so it is natural first to try to identify matter as confined wave energy of these types. Say that one of these waveforms, or a mixture of them, are confined in a certain area of space from where they are not allowed to escape. We can think of it as a weight-less box with reflecting walls embedded in the spatial continuum wherein wave energy is bouncing around. It can be shown that con-fined wave energy of any kind will exert a pressure that amounts to one-third of the energy density on whatever keeps it at bay (see e.g. Sec. 13.3). The pressure will displace an amount of spatial mass that is independent of how densely the energy is packed in the box (see Fig. 5.1).

A pattern of confined energy would meet some of the properties of matter. It could move in space and would possess momentum. To accelerate it, we would have to add a component to the energy flow in the direction of the acceleration and thus increase the total amount of energy and accordingly the mass. Great or small masses, like planets in the sky, or smaller bodies on Earth, can with a high degree of accuracy be described by Newton's[1] laws of motion, but at great velocities compared to the speed of light, his laws seems to be insufficient because mass increases with speed. Newton, however, formulated his second law of motion by stating that the *change of motion* (not the change of velocity as we now tend to formulate it) is proportional to the force that acts upon a body.[2] It is impressive that he intentional or unintentional formulated his law this way, be-cause if we interpret motion as momentum, and combine it with the knowledge of the 20th century that matter and energy are equiva-lent properties so that the energy that accelerates a body goes as an addition to the body's rest energy, Newton's second law of motion would in fact be relativistic (see Fig. 5.2).

Even if the idea that matter entirely consists of confined wave energy meets some of the properties of matter, it immediately en-counters several serious objections. How can wave energy be con-

---

[1] *Isaac Newton (1643-1727)*

[2] He writes in his Principia: "DEFINITION II: The quantity of Motion is the measure of the same, arising from the velocity and quantity of matter con-junctly". "Motion" is clearly not meant to be simply velocity!

Newton's second law: $F = \frac{d(mv)}{dt}$,   where $v = \frac{ds}{dt}$.

Einstein's energy equation: $E = mc^2$.

From these equations we have:

$$dE = \frac{d(mv)}{dt}ds = \frac{d}{dt}\left(\frac{E}{c^2}v\right)ds = \frac{1}{c^2}\left(v \cdot dE + E \cdot dv\right)dv,$$

$$\frac{dE}{E} = \frac{1}{c^2}\frac{v \cdot dv}{1 - v^2/c^2},$$

$$\ln E = \ln\frac{1}{\sqrt{1 - v^2/c^2}} + \ln C = \ln\frac{C}{\sqrt{1 - v^2/c^2}},$$

$$E = \frac{E_0}{\sqrt{1 - v^2/c^2}},$$

$$m = \frac{m_0}{\sqrt{1 - v^2/c^2}}.$$

NEWTON'S SECOND LAW is relativistic if we combine it with the knowledge of the 20th century that matter and energy are equivalent properties. The change in energy is like the force times the distance over which the force is acting so that the resulting energy of the system is the rest energy plus the energy added by the accelerating force.

Figure 5.2: Newton's second law.

fined inside small volumes? That is forbidden by Huygens' principle, which states that every point on a wave front may be considered to be a new source of disturbance from which spherical wavelets issue. Thus ordinary waves in a continuum will always spread in space and die out. Elementary particles like electrons are known to be point-like, or very nearly so. How then can a wave packet be point-like? Confined energy with the density found in a material particle is so huge that space itself would be ripped into pieces if local stretching forces should keep the energy at bay. An even more basic difficulty arises. Waves in an elastic continuum is, as we have seen, either solenoidal or irrotational, the former propagating with approximately the double of the speed of the latter. These waves are literally independent of each other, but both effects are probably needed in order to create a model of matter. These and other questions I will take up in later parts of the book, but for now I'll conclude that matter essentially is composed of confined disturbance energy in the spatial continuum, and that there – at least in some connections – is a one to one dependency between energy and displaced spatial mass.

5.2   Reflecting singularities

Waves in an elastic continuum can, as we have seen, be of two very different types, namely longitudinal or transversal waves. Longitudinal waves consist of oscillation between compression and rarefaction, and transversal waves are shear waves characterized by alternating rotational components transverse to the direction of the wave movement. The two wave forms are literally independent of each other when the deformations are small, but not necessarily so with greater deformations that might occur if waves should come to smash together into a singularity.

The single oscillating node.

Waves in the spatial continuum can in principle be created by different thought experiments. If an imaginary rigid plane surface is rocked back and forth normal to the plane, it will create a plane compression wave, a P-wave. If it instead is set to rock up and down along a line in the plane itself, it will create a plane shear wave, a S-wave. By a similar thought experiment we could create a spherical compression wave by inserting a small pulsating sphere into the spatial continuum. Equation 6.5 on page 109 is a solution for such a wave, and it shows that the wave equation has two solutions; one for waves moving outwards from the source, and one for waves moving inwards towards the center of the spherical wave front.

Let us extend the thought experiment by enclosing the space around the pulsating sphere with a concentric rigid shell. The wave generated by the pulsating sphere would then be reflected by this outer wall and turned into an inwards moving spherical wave. Say that we in the meantime gently remove the inner sphere and let the wave clash into a singularity at the center. It can be shown that the wave would be reflected back to an outwards moving spherical wave. Hence the wave – when first created – would bounce back and forth between the rigid shell and the center in an eternal round dance (see Sec. 6.2 on page 111). By adjusting the relation between the wavelength and the radius of the shell, the inwards and the outwards moving part of the wave train can be made to interfere such that a standing wave is formed inside the shell. Further by enlarging the radius of the shell in steps of half wavelengths and letting the innermost amplitude remain unchanged, more and more energy will go into the oscillations. Since it only is a thought experiment, we could even let the radius and hence the energy grow towards infinity, or nearly so, only limited by the size of the entire Universe. Crazy as it might seem, it could be considered to be a point-like elementary particle filling up the whole Universe and containing all the energy there is! In the rest of this book, however, I shall treat all oscillating systems as superposition of oscillating nodes of the type outlined above. If the total energy approaches infinity, then the system is forbidden. But if I can find constellations of nodes where destructive interference limits the total energy to a finite level, I'll assume that such systems may be stable.

So far I have only discussed pressure waves, but much of the same mechanism can also apply to shear waves. We could create a spherical shear wave by letting the sphere in our thought experiment rotate clockwise and counter clockwise around an axis through the center of the sphere. The shear waves would spread in space and take the shape of standing waves between the rigid concentric shell and the singularity at the center in much the same way as the compression waves. The rotation can be given a direction in space according to the right hand rule. Imagine that you take a grip with your right hand around the rotating sphere, and let the fingers point in the direction of the movement. Then positive rotation can be defined as pointing in the direction of your thumb. Thus an oscillating node can be defined as a node that rotate alternatively between, say the right and left direction.

In the next sections I shall try to discuss if multitudes of oscillating nodes of the two types outlined above, or combinations of them, can be realized in the Universe, and thus make up systems of energy concentrations that can make up matter as we see it around us. It is already an interesting thought that the basic building blocks of matter might be oscillating nodes with an infinite amounts of energy. They have a distinct location, but yet they extend throughout the whole Universe. In a sense we are led towards accepting the *Mach principle* that the whole Universe matters locally, and the principle of *non locality* of events.

5.3 Myriads of oscillating nodes

As a result of the Big Bang, space gets filled with myriads of oscillating nodes of the two types – or combinations of them – as outlined above. In a space of infinite extension, the nodes would act like sources that emit energy into space and die out. The space, however, is not infinite, but surrounded by a reflecting wall at its border so the disturbance energy inside is kept at bay. The result may be that energy, which is emitted from one node, finds its way into one or several nearby nodes where it is reflected and so on in an eternal round-dance. In this way the Big Bang leaves a compressed spatial continuum filled with nodes oscillating with different frequencies throughout all of its extension. It is very much similar to an amount of ordinary matter heated to a certain extent and surrounded by an insulating surface like a liquid in a vacuum flask. Another consequence of this picture might be that when a node first is created, it would last forever, and like heat, is connected to the kinetic energy of distinct particles, the spatial heat is connected to the oscillatory energy around each node. Hence empty space is not empty at all, but filled with energy, basically consisting of waves that move with the propagating speed of transversal and longitudinal waves. A soup of oscillating nodes as suggested above could well be the chaotic plasma of the very early Universe from which the ordered system of material entities were formed when the Universe was cooling down to the about three degrees Kelvin we see in the cosmic microwave background (CMB) today.

I finish this section by stating that in search of the basic building blocks of matter, we have got to focus on studying the two kinds of oscillating nodes and the interaction between them. It may seem strange that the basic building blocks should have an infinite amount of energy, but if constellations of them should happen to have a finite energy, then breaking up such a constellation would require a vast amount of energy, and therefore the force that keeps such a system together, would be extremely strong.

A doublet of oscillating nodes

In the first fraction of a second after the Big Bang all possible states of singularities would occur at the same time. Say that an expanding node happened to be near an equally strong collapsing node. Then spatial mass from the exploding node could find its way into the imploding node. A short time later the pressure in the exploding node would have been turned into a vacuum and vice versa for the pressure in the imploding node. Thus the roles have been reversed and so spatial mass has started to oscillate between the two entities.

The field around the group can be seen as a superposition of the fields around each of the nodes. Since they are oscillating in opposite phase the interference will be mostly destructive, so the total field energy will almost certainly be lower than if the nodes were seen as two separate objects. To see what force there is acting between two nodes of this type, we could think of them as small pulsating spheres immersed in a compressible liquid.

The effect of attraction and repulsion between pulsating spheres immersed in either an incompressible or a compressible fluid were discussed in the last part of the nineteenth century, mainly as a follow-up of Bjerknes'[3] proposal that gravitation could be explained as attraction between pulsating spheres with equal phase and frequency immersed in an incompressible fluid. A. H. Leahy, however, pointed out that the effect was reversed in the case of compressible fluids if the distance between the pulsating spheres exceeds half a wavelength[4].

It was found that in the case of pulsating spheres in a compressible fluid, the effect of two spheres pulsating in opposite phase and the same frequency would attract if the distance is greater than half a wavelength and repel if it is shorter. So the most stable distance between two nodes oscillating with the same frequency, but in opposite phase, is half a wavelength. If the distance is shorter, they will repel, and if it is longer, they will attract. However, if we add the the total field energy around the doublet, we would still get an infinite amount of energy, so the doublet would not be stable. It would quickly radiate away all its energy (see Section 6.4).

Strings of oscillating nodes

The next case I'll discuss is a string of oscillating nodes, one node oscillating in opposite phase to its two adjacent nodes. As with the doublet, the distance between the nodes has got to be half a wavelength. At that distance the force between the nodes shifts from repulsion to attraction.

Again, think of the string as a series of pulsating spheres each pulsating in opposite phase to its neighbours. When the separation is half a wavelength the force between them is zero. That means that the total energy in the system has got to have a minimum at that separation. We could say that there is a potential dip, or potential well, at that point. The depth of the potential well determines how strong the force to keep up this distance is, i.e. the strength of the string. The over all force between the objects has been compared

---

[3]Carl Anton Bjerknes, mathematician and physicist from Oslo, Norway (1825-1903).

[4]Transactions Cambridge Philosophical Society XIV (1864) p. 45

with gravitational forces between material bodies, so it follows an inverse square law. Hence the repulsion between the items should increase towards infinity when the distance approaches zero, and the attraction should still be high at a distance that exceeds half a wavelength, so the potential well should be rather deep. It follows that a string of this type is extremely strong.

To find the total energy of such a construct, we would have to add up the field energy in the surroundings of the string. First we find that the field a short distance to the side of the string has got to be almost zero, because seen from some distance away, each pulsating sphere has a neighbour that is pulsating in opposite phase. The field will approach zero because of destructive interference. Otherwise, we see that the interference along the string is mostly constructive. Thus we see that the field energy will be concentrated in a relatively narrow tube along the chain of nodes I assume that the energy density along an infinitely long string will be finite, so the total energy will have a finite limit per unit length. Possibly need the string not be straight, but can perhaps even form a closed curve with a finite energy.

Let us take a look back at the Big Bang. Initially space was filled with a huge amount of oscillating nodes, some oscillating between compression and rarefaction and some between opposite directed rotation. If it is so that the energy needed to sustain the oscillation has its minimum when the nodes are organized along strings, then the myriads of nodes would soon organize in that way, and space would almost immediately become filled with a vipers nest of strings of oscillating nodes.

Dark matter?

The first tendency towards the formation of an ordered system after the chaos of the Big Bang, is – as we have seen – supposed to be oscillating nodes tending to organize into long chains of nodes oscillating with different amplitudes and frequencies. In such strings of nodes, the spatial mass is in equilibrium, what is displaced from half of the nodes can be found in the other half. But if other constellations are possible, they certainly will occur. Anything that can happen, will happen, but with different probabilities. In the first fraction of a second after the Big Bang, however, every possible pattern will probably immediately be broken down by impacts from other systems, but as the Universe cools down, more and more patterns can become stable. So, let us try to figure out possible systems that may evolve.

The Big Crunch is thought to have been an inwards moving tide of spatial mass, that tended to end up in one gigantic singularity. But before it came that far, small irregularities broke down the symmetry so the Big Crunch ended up, not in one singularity, but in a boiling inferno of small singularities, i.e. nodes, that each exploded and sent spatial mass away from the site leaving a vacuum in its wake. An evacuated node, let us call it a bubble, might have been filled up again with spatial mass from another exploding node, which took over the role as a new bubble and so on.

Say that one such bubble happened to occur in a chain of nodes. It may then have been filled up with spatial mass from its two adjacent nodes, and thus been split into two smaller bubbles moving in opposite directions. Such bubbles can be compared with a weightless object, e.g. a sphere, moving through a perfect fluid with no turbulence. *D'Alembert's paradox* states that the resistance is zero when the sphere is moving with uniform velocity through a perfect fluid. If we increase the speed, however, the fluid around the sphere also changes the velocity state which increases the total kinetic energy in the liquid. In spite of the non-resistance movement through the liquid, the weightless sphere apparently has obtained a momentum that happens to correspond to one half of the mass displaced by the sphere. Therefore a moving node that displaces some of the spatial mass, has a momentum that corresponds to half of the displaced spatial mass [9, volume II, page 162]. Thus, in order to conserve the momentum, the two bubbles will continue to move in a stepwise manner away from each other.

Let us study a moving bubble, in some more detail. In a chain of oscillating nodes the spatial mass that is displaced from half of the nodes typically goes to compress the domain around the rest of the nodes. Hence the chain of nodes per se does not displace any amount of spatial mass at all. But as we have seen, there may be a single node in the chain that is evacuated to a higher degree than what is taken up by the other nodes in the chain. A node of this kind is for convenience said to be a bubble. The bubble is not stable, but will immediately be filled up with an inward stream of spatial mass from the next node in the chain. In this way the bubble is moving in a stepwise manner from node to node along the chain. When the leftover pressure dump is exactly filled up, the inward stream of spatial mass has its maximum speed and will continue for a while until the compression peak stops the movement and reverses the stream. Therefore nodes that are left behind have got to oscillate with decreasing amplitudes as the bubble recedes.

On the other hand the next node in the chain cannot suddenly be blown up to the exited state, but has got to start oscillating with increasing amplitude in good time before the bubble arrives. Hence the arrival of the exited node has got to be preceded by a pilot chain of nodes that oscillate with increasing strength as the inflated node approaches. The entire chain of nodes may extend with decreasing amplitudes far out – in principle an infinitely long stretch – in both directions. This picture conforms well with the findings above where I found that oscillating nodes only are possible in an infinitely long chain of oscillating nodes. Here the assumption has got to be modified to incorporate very long string of nodes with steadily increasing strength, which I'll dub the *preamble* that culminates in a strongly evacuated node, a bubble, whereupon the trailing nodes, which I will dub the *postamble*, gradually decrease towards zero strength – or perhaps more correct – towards the background noise of the universe (see Fig. 5.3).

There are no reasons why a highly compressed node cannot move along in the same way as a bubble. If the bubble represent a moving particle, then a compressed node could represent its antiparticle, and if two such particles meet, they would probably annihilate and turn into some other configurations. The two entities sketched above resemble in many ways two elementary material particles.

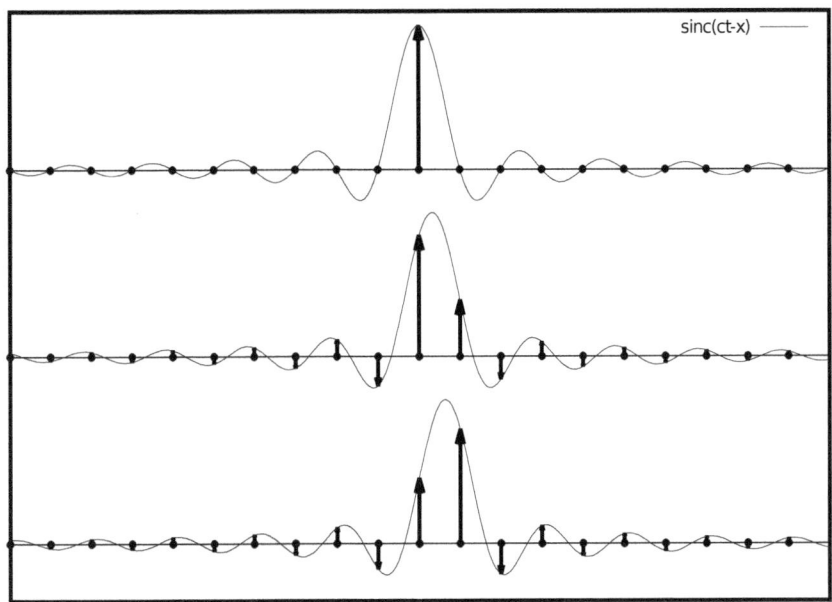

Figure 5.3: Chain of nodes.
A chain of nodes alternating between compression (down arrows) and rarefaction (up arrows) culminating in a highly evacuated node, a bubble, whereupon the nodes in the postamble fades down towards zero. The three graphs show the movement towards the right in three successive steps.

They have energy and momentum, and at the same time wavelike and point-like properties – wavelike because the pre- and postamble alternates between compression and dilatation, and point-like because of the highly exited node that is carried along within the entire chain of nodes. The nodes themselves are stationary. They only oscillates with changing amplitude while the exited node moves forwards by jumping from node to node along the chain.

The entities above may or may not be realizable, but if they are, they might be some sort of dark matter. Real elementary particles are known to have spin, and the possible particles above cannot fit into that pattern. If dark matter is an accumulation of particles of this type, it is hard to tell how they can be bound together into bodies. Perhaps they can only interact in some scattering processes and exhibit some plasma-like structure that can only gather into greater structures because of gravitation. Next, they will be extremely difficult to observe with any kind of equipment based on detecting S-waves, which in this model is similar to electromagnetic waves.

5.4   The photon

Most of what is said about nodes that oscillates between compression and rarefaction, can be said about right and left spinning nodes. But there are also major differences. While positive and negative pressure can be expressed as a scalar, rotation can only be expressed as a vector. The displacement components around a rotating node, are always in planes normal to the direction of rotation. There are no components along the rotation axis.

For a chain of right and left spinning nodes to appear, it is an absolute condition that all the axes of the spinning nodes in the chain are pointing in the same direction. Only then can the interference abeam of the chain be destructive and the chain obtain a finite energy per unit length. This restriction impose an interesting property to deformations in the spatial continuum. Say that all the axes in a chain are pointing in the z direction, then the displacement vectors, **u**, only have components in the x-y plane. This makes it possible to describe the displacements as the real and imaginary part of a complex number, which greatly simplifies the treatment of the displacement field around such a string of nodes.

A string of right and left spinning nodes making up a standing wave, can be described as two identical progressive waves moving along the string in opposite directions with the characteristic speed of solenoidal waves. The superposition of the displacements adds up to a standing wave. The rotation in each node will alternate between the right and left direction, so no permanent twist will be applied to any of the nodes, and the sum of the spin components is zero.

Consider a chain of right an left spinning nodes where say the right spinning nodes in the preamble successively build up to a highly right directed twist in the excited node and then decrease towards zero in the postamble. The next node in the chain follows the same pattern and reaches its peak value, also to the right, half a period later. In this way the right rotation jumps from node to node along the chain in the same way as the bubble did in the previous example. The pattern is much the same as in Fig. 5.3, but here the up and down pointing arrows in the graph are to be interpreted as the spin direction with the up arrows directed toward the right.

We could compare the deformation field around a single oscillating node with the electromagnetic field around a dipole antenna. The electric field would correspond to minus the velocity field, and the magnetic field to the rotational field. A half wave dipole antenna is somewhat directional with no radiation along the direction of the antenna, but with sharply increasing radiation in all other directions. If we place some shorter elements in parallel with the dipole in front of it, and a somewhat longer element behind the dipole, the antenna becomes increasingly directional. Such an antenna is known as a Yagi-Uda antenna, or just Yagi antenna. In much the same way as with a Yagi antenna the nodes in a chain of nodes is assumed to concentrate the deformation field to a comparatively narrow tube along the chain. I'll also assume that a stretching of the chain, or a small deviation in the direction of the axes of rotation will lead to a sharp increase in the field energy and therefore make such a string almost unbreakable.

If we take a closer look at a chain of nodes that builds up to a highly excited node, we see that the pattern with every other node pointing to the right is broken when the excited node reaches its maximum. This causes the nodes of the preamble to be one half cycle out of phase with those in the postamble (see upper graph in Fig. 5.4). This should be corrected in some way, and the most obvious way is that the spin direction flips over from right to left as the rotation reaches its maximum. Then the rotational directions in the pre- and postamble will be in phase (see lower graph in Fig.

5.4). I shall assume that this is precisely what happens each time a new node reaches its maximum rotation.

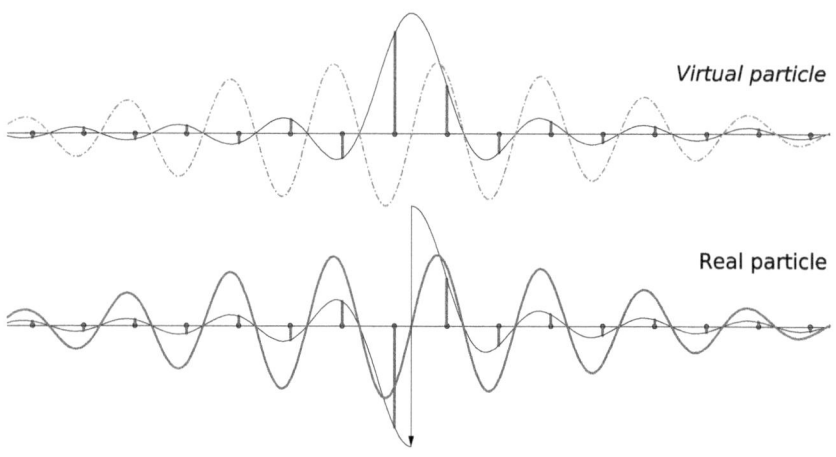

Figure 5.4: Particle waves.
The upper graph shows that the wave generated in the preamble of a particle moving from left to right is out of phase with the oscillations in the postamble. If the torsion, black graph, flips over from the right to the left direction in the exited node, however, the oscillations in the whole system will be in phase (lower figure), and a following progressive wave may be generated (sinusoidal graph).

The right and left spinning nodes in the pre- and postamble produce up and down directed displacements in between the nodes, that may trigger a plane transversal wave in the vicinity of the chain. The wave will follows the system as it moves along with the characteristic speed of solenoidal waves. Without the flip over, the wave generated in the preamble would be half a period out of phase with that generated in the postamble and thus tending to suppress the formation of a following wave. This will be a contributing factor to provoke the flip over.

The flip over of rotation from right to left is a dynamic process that cannot happen without some kind of spin is involved, and since this event is repeated every half period of the oscillation, the spin appears to be a property of an entity of this kind. This raises a new question: Is it at all possible to have an ongoing spin in an elastic continuum? That a rotation can be performed without imposing a twist to a body, is best demonstrated by the Balinese cup trick. The Balinese dancer takes a cup, partly filled with water, and turns it a full 720 ° around, or a multiple thereof, without loosening the grip of the cup, or turning herself around, and of cause, not spilling a drop of water! Similar tricks are known as Feynman's plate trick, Dirac's belt trick, spinor spanner, Bredon high-five, or quaternionic handshake, and they are all demonstrations of the mathematical property of *spinors*. It is noteworthy that these tricks involves that a volume element turns two full revolutions and never only 360 °.

In this model the spinor serves the purpose of bringing spin into the spatial continuum without imposing a permanent deformation. Spinors are notoriously difficult to treat mathematically, so I shall not try to go deeper into that subject, but only assume that flip over of spin as sketched above is possible, and that it is related to the concept of spinors. This will be the key to understanding the last property of a photon, namely its intrinsic spin.

A full picture of a photon is starting to evolve (see Fig. 5.5). The excited node makes it a particle-like entity, which can only be created and absorbed as a unity, e.g. accounting for the photoelectric effects. Its associated progressive wave accounts for its wave-like property. The chain of nodes of a photon can for example only pass through one of the slits in a two-slit experiment, but the associated wave will travel through both slits and interfere in a constructive or destructive manner behind the screen, and thus guide the string with different probabilities in different directions. The mechanics outlined above will also explain why a photon has spin. The flip-over of rotation is thought to be 'wrenched' through the direction of movement where it performs like a spinning object, namely a spinor. Perhaps we even can say something about the strength of the spin in relation to the energy, see Fig. 5.6.

And finally, like any wave movement the photon model has a momentum corresponding to the progressive wave, but it does not displace any spatial mass and thus do not have an additional mo-

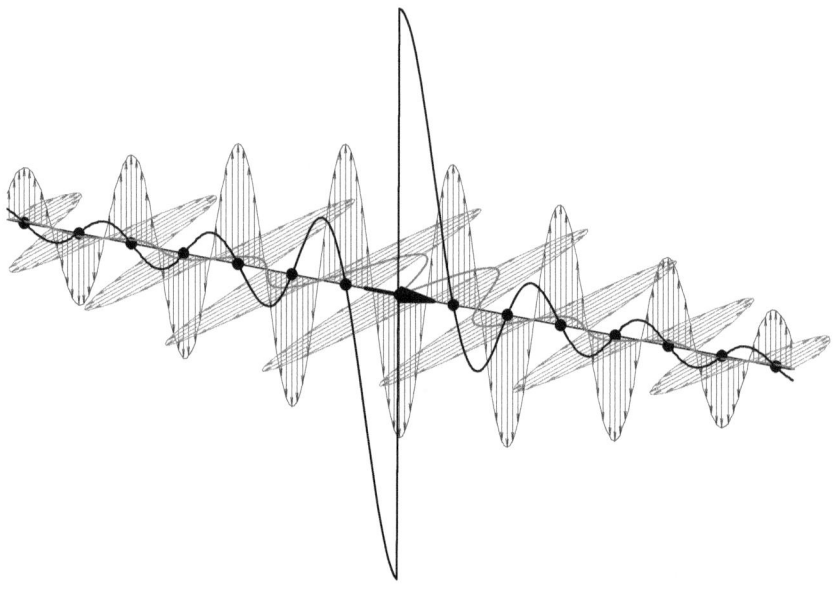

Figure 5.5: The photon.
A string of up and down rotated nodes (black dots) in the preamble ends up in a strongly excited node where it flips direction from up to down and thereupon decreases towards zero in the postamble (black graph). The flip-over creates a spinor in the forward or backwards direction (black arrow). A progressive transversal wave is generated in the surroundings (sinusoidal graph).

mentum like that with a moving bubble. Hence it can be said to be massless.

5.5   Coupled oscillations

In the Linear Theory of Elasticity only small, almost infinitesimal, deformations are considered, and since solenoidal and irrotational waves in that theory are independent of each other, there can be no coupling between them. In the near vicinity of the oscillating nodes, however, the deformations can have great gradients, and

From elementary textbooks we find that a spinning body has an energy given by $E_s = 1/2 \cdot I\omega^2$, where I is the momentum of inertia and ω the angular frequency. Its spin is given by $L = I\omega$. We obtain a relation between the spin energy and the spin given by:

$$E_s = \tfrac{1}{2}I\omega^2 = \tfrac{1}{2}L\omega.$$

The energy in a chain of oscillating nodes is thought to split evenly between potential and kinetic energy. Hence the total energy of a photon is:

$$E_p = L_p\omega.$$

If we assume that the energy of a photon is given by:

$$E_p = \hbar\omega,$$

then the intrinsic spin of the photon is like:

$$L_p = \hbar.$$

Figure 5.6: A spinning body.

the Navier-Cauchy equation no longer is sufficient to fully describe the deformations. The reason is that the partial derivative of the particle speed with respect on time no longer gives the correct acceleration. We see this effect best in fluid dynamics[5]. A Venturi tube is a tube with a narrow passage where a steady stream of an incompressible fluid is forced to increase its speed. In a steady flow through such a tube the partial time derivative of any part of the fluid is zero, while a volume element has an increased velocity, i.e. an acceleration, as it passes from a wider to a narrower part of the tube. A simple laboratory experiment shows that the pressure in the fluid falls when the velocity increases.

---

[5]In fluid dynamics we have got to modify acceleration to: $\ddot{\mathbf{u}} = \partial\mathbf{v}/\partial t + (\mathbf{v} \cdot \nabla)\mathbf{v}$.

Consider a node oscillating between compression and rarefaction. When the flow is at its maximum speed, the partial time derivative of the speed of, say an inwards flow, is zero. If we follow a volume element of the spatial continuum, however, we see that it passes through a diminishing surface and thus is forced to increase its speed. Although the partial derivative is zero, the volume element has an acceleration ($d\mathbf{v}/dt = \dot{\mathbf{v}} \neq \partial\mathbf{v}/\partial t$). In this environment the two characteristic waveforms in the spatial continuum no longer are independent of each other, and there may be a transfer of energy from compressional to rotational energy and vice versa.

It is an established fact in fluid mechanics that in the absence of external forces, spin in a perfect fluid is a conserved property, so it seems likely to expect that spin in the spatial continuum is conserved. It can be a bit tricky to see how this can apply to an elastic continuum, but as a first step it is important to distinguish between spin and rotation. Spin is a conserved property, but rotation is not. Let us look into a practical example. When an ice dancer with her skating skills sets her body into a slow rotation, she gains a certain amount of spin with parts of her body as far as possible from the rotational axis. Then she straightens out her body and pulls her arms and legs as close as possible to the rotational axis, and then – almost miraculously – her body goes into a fast rotation that might deceive us to believing that the spin has increased. But that is not so. A closer examination shows us that spin has not changed. Rotation, however, is faster and the rotational energy is greater. But then, where does the energy come from? When the dancer pulls her body parts towards the axis she acts against the centrifugal forces and that is where she puts more energy into rotation and makes her beautiful pirouette. When she – now acting in the same direction as the centrifugal forces – spreads out her body parts away from the rotational axis, the rotation almost entirely stops, but to also stop the spin she has got to scrape the skates against the ice. Hence rotation and rotational energy can change while spin is conserved. It is therefore no violation of any fundamental principles that energy stemming from irrotational movements are transformed into rotational energy and vice versa as long as spin is conserved.

We clearly see the resemblance with movement of spatial mass to and from a node. When spatial mass is moving inwards towards the singularity, it may give up some of its kinetic energy to rotational energy and gain it back again when the movement is reversed. This exchange of energy is a necessary part of a coupling between solenoidal and irrotational energy. Like the ice dancer had to initiate the rotation with a certain amount of spin, the oscillating nodes have got to have at least a small element of spin to initialize the rotation. However, if we add all the spin components in the pre- and postamble the net spin is zero because of symmetry. The spin carried by the particle, is entirely a result of the flip over of spin from one direction to the opposite direction. It is likely that the effect is that of a spinor.

5.6 The material body

In the preceding section we saw that there under certain conditions may be a coupling between solenoidal and irrotational wave movement to and from a node. A string of nodes that in the preamble builds up to a highly excited node with both compression/dilatation and up/down rotation, and where the rotation flips over from the up to the down direction, followed by a postamble where the deformations decrease towards nil, may be the basic elements from which matter is built. A detailed picture of such movements may reveal different possibilities that account for the multitude of states that are necessary parts of matter, like quarks and leptons.

Let us take a review of known properties of material bodies and try to imagine how entities as sketched above should behave in order to fit into the picture. First we know that the speed of a material particle cannot exceed the speed of light. Thus the exited node should move along the string with this speed, while the string itself is internally wound up in some way such that the net movement of the particle is with some speed less than the speed of light. So when I in this context speak of an elementary material particle, I shall mean an entity as sketched above moving with some velocity in space. Its energy should be given by the oscillatory frequency times a constant, h, like Plank's constant[6]. The energy varies according to the particle's speed, so two identical particles may have different oscillatory frequencies depending on their state of movement. If two identical particles were oscillating with the same frequency, the oscillating nodes would come in conflict with each other, so this should be forbidden. Hence particles, at least some of them, should obey the Pauli exclusion principle. We can draw this assumption a step further and forbid that the peak moment for two different nodes should coincide. Then, since all elementary particles have their own unique oscillatory frequency, the frequency of such peak values as seen from the outside world, has got to be like the sum of all these separate frequencies.

Basically then, in this model a material body consists of several groups of oscillating nodes, each group oscillating with its own unique frequency, and carrying with it an energy given by the frequency times Plank's constant, h. In each group the exited node reaches its peak value independent of all the other groups, so seen from the outside world, the frequency of all these events becomes the sum of the individual frequencies. Hence the energy of the whole body amounts to this sum times Plank's constant. Finally, if we assume that all the nodes in a body have parallel rotational axes, then all the displacement vector components will be in planes orthogonal to these axes. Thus the displacement field can be expressed as a superposition of only two components in that plane, which makes it possible to describe displacements with complex numbers.

5.7   Electric charges

In Chapter 2 I discussed how electromagnetic fields could be represented in a mechanical model of space and matter, and in Sec. 2.2 I found that the mechanical counterpart to an electric field has got to be the negative of a velocity field. From electrodynamics

---

[6]In Chap. 5 I have tried to see the relation between frequency and energy, and found that in this model such a relation indeed is possible.

we know that electric field lines goes from a positive electric charge to a negative charge, but since the electric field was defined as the negative directed velocity field, the velocity field lines have got to go from a negative to a positive charge. Hence it is a consequence of this model that the electron has got to act like a source and a positron like a sink (see Sec. 2.5).

At first, it almost looks like a paradox that there may be a velocity field between two point-like objects like a source and a sink. It seems to require a mysterious loophole between the two entities through which spatial mass can flow back from the sink to the source. Loopholes, however, have no place in this model of space and matter, so we must look for other possibilities.

In a section above I found that strings of nodes that oscillates between compression and rarefaction can bring with them both holes in the form of evacuated bubbles and chunks of spatial mass in the form of highly compressed nodes, while *real photons* do not do any such thing. But suppose that *virtual photons* can bring with them either a compressed or an evacuated node, then such photons can transport spatial mass between sinks and sources. Say that a virtual photon carrying a surplus of spatial mass from a point A to a nearby point B before it disappears, and call it an up polarized virtual photon, or just an up photon. The effect of such an event would be that B has received some amount of spatial mass that has got to be transported back to A. Then B acts like a source and A like a sink. The virtual photon performs like a loophole between the two entities. A virtual photon carrying a hole – call it a down polarized virtual photon, or just a down photon – would act in a reverse manner.

Now, let space be filled with events where up and down photons are created by pair production and move for a short distance until they meet an opposite polarized virtual photon and then annihilate each other. The up photons will drag spatial mass in the same direction as they themselves are moving while the down photons will push spatial mass in the opposite direction of their own movement. Since the up and down polarized virtual photons are moving in opposite directions, they will set up a velocity field that can be interpreted as an electric field. Everything needed to maintain such a field, would be two engines, one that emits up and one that emits down photons.

5.8 The electron

In a previous section I proposed a model of a photon. Here I will try to find out what an electron might prove to be. We know that a photon with sufficient energy when passing near the strong field of a nucleus, can split into two leptons, a positron and an electron. Therefore an electron in many ways should resemble a photon, but there are some major differences that have got to be accounted for.

A photon moves along with the speed of light which is synonymous with the speed of solenoidal waves. The two leptons, however, may be moving with any speed lower than that. Therefore, when the proposed model of a photon splits up into two new entities, the two chains of nodes representing the two new objects, have got to curl up in some way such that even if the exited nodes still moves along the chain with the speed of light, the net speed may be anything lower than that. A photon has integral spin, but if spin is a conserved property, then the two new entities can only have half of that spin each. Perhaps that is the reason why they do not move in straight lines, but twirl away along some irregular paths.

We have already seen that in this model the property of electric charge is related to sinks and sources. Charge is a conserved property, and since the photon model has no sink/source properties, sinks and sources have got to be created by pair production – one sink for every equally strong source. That will guarantee for the conservation of charge in the present model of space and matter.

Suppose that an electron works like this: Basically it is a curled up string of nodes with a preamble, an exited state, and a postamble. A node in the preamble has a solenoidal oscillation between the right and left direction coupled to an irrotational oscillation between compression and rarefaction. Say that rotation in the direction that eventually is going to flip over, is coinciding with compression, and that both rotation and compression flip over to the opposite extreme when they reach their exited state. In order to build up compression, the mother particle has got to suck in spatial mass from the surroundings along its preamble. Because of the flip over of compression, the nodes in the postamble are evacuated to some degree and have got to be filled up as they fade down to the surrounding pressure. Thus the mother particle has got to suck in spatial mass along its entire length and it would act like a sink. If the proposed flip-over of compression really is going to take place, the mother particle has got to suddenly get rid of a chunk of spatial matter when it reaches its peak value. It could do that by emitting an up polarized virtual photon each time it reaches that state, or the flip over could be initiated by colliding with a bubble to be filled up wholly or partly, then with some residual particles in the wake. A positron is supposed to act in a similar way. Together they would be the engines that produce electrical charges and keep up the electric field between them.

This picture of an electron seems to fit into the conception that, although an electron acts like a particle when interacting with other particles, it behaves like a smeared out pattern when treated as a wave. I imagine that the string, which the exited node has got to follow, can wind up in smaller or greater loops that may pass by almost any places in space. Since the excited node has got to strictly follow the chain, it can also be found almost anywhere with varying probabilities. It is probably impossible to spot the position of the particle itself at a given time, but in this model such a position certainly exists. In a theory developed by the late physicist David Böhm, he called this position a "hidden variable" (*A Suggested Interpretation of the Quantum Theory in Terms of "Hidden" Variables. II.* Phys. Rev. 85, 180 – Published 15 January 1952 by David Böhm.) The modern view is to describe a probability amplitude, i.e. the probability to finding a particle in a given volume element. If we extend the volume element to encompass the whole universe, the probability is one, because the particle has got to be somewhere, but in smaller volume elements, the probability is something between zero and one.

The last property of the electron I shall mention briefly is how it behaves in an electromagnetic field. If we simply accept that electrons behave like sources, then it follows directly from the Navier-Cauchy equation that a density of such items behaves exactly like a density of charges (see Fig. 2.7). For example a stream of electrons in a conductor creates a magnetic field around the conductor.

5.9   Strong forces, quarks and gluons

In this model a positive charge is an engine that produces up polarized virtual photons and receives down polarized photons. It acts like a source. A crucial point is that the virtual photons are free. If we increase the distance between the charges, the only thing that happens is that more virtual photons are under way at the same time. The energy that goes to increase the field between a positive and a negative charge leads to Coulomb's law.

Now consider particles that are so close together that the connecting virtual particles are hooked up in both ends to the sender and receiver. Then the transported amount/deficit of spatial mass has got to jump from node to node until it is absorbed by the other particle. The only way to increase the distance between the particles would be to insert additional links in the chain of nodes between them. To remove the particles from each other, would imply that the chains have got to be broken. If the chains are very strong, which is assumed in a section above, the force that binds such particles together should be strong enough to account for the *strong force* of nature. Then the outlined particles would be *quarks*, and the connecting strings would be *gluons*. In the following I'll try to put some more meat on the bones to see if this conception can be fitted into what we know about quarks and gluons.

Consider two particles, a particle and its antiparticle, so close to each other that the virtual particles emitted from each of them just are long enough to hook directly up with the other. The particle and the antiparticle have got to reach their excited states in succession, so the length of the connection chains have got to adjust to a specific length so that the surplus/deficit of spatial mass from one particle reaches the other just in time to hook up when that particle reaches its peak value. Then we have two particles that are tied to each other by two unbroken chains of nodes, and we call such a the chain a *gluon*. Even if the connection imply an electrical field, the strength of the gluon is far stronger than the electric forces. Hence the particles are tied together by a force that is much stronger than the electric field, and therefore it is called a *strong force*. Such a doublet of particles are known to belong to a group of particles called *mesons*, and the two particles it consists of, are called *quarks*.

If the time between two excitations for one of the particles is organized along a full circle, we see that the excitations for each particle is 180° out of phase. Colour is often organised in the same way, and then the two particles can be said to have a colour and its anti colour. The transported mass or bubble are exchanged at different times, so the tip and the tail of the connecting string can be ascribed both a colour and an anti colour. Thus a *colour charge* in this model can be interpreted as a phase difference. Because the two particles are kept from getting into the same phase, they can exist alongside each other without being annihilated. It is also the reason why they are not breaking the *Pauli exclusion principle*.

A meson is not stable so the direct connections between the two quarks are easily broken and they fly apart with a speed too great for the electric attraction to stop them. Quarks, however, cannot exist in isolation, so the meson disintegrates into some other real particles.

Consider two positively and two negatively charged particles, and let three of them be connected by direct connection and the third be an electron connected to the group by free virtual particles. Recall that the connections are entirely performed by exchange of spatial mass between the particles. By a direct connection there is an unbroken chain of nodes between the particles along which the surplus/deficit of mass is transported, while free particles carry with them the surplus/deficit of mass. In the first case it is the strength of the string that is dominating, while in the second case there is only a changing number of free virtual particles. In the sketched set-up the three particles are closely bound together in a particles, known as a *proton*, while the electron is found orders of magnitudes farther away from the proton. The construction is of course the most abundant matter in the universe, namely hydrogen.

One of the basic properties in this model of space and matter, is that the excitation points for all the separate entities are quite independent of each other and cannot coincide in time. When studying the proton it is important to be aware of that. The three particles emit and receive the surplus and deficit of spatial masses in succession, and we can organize the time for one full period along a circle in the same way as we are used to organize the colour. Hence we can ascribe a colour to each event and call it a *colour charge*. For example a bubble emitted from one particle will need some time to travel to the next node and arrive there just in time to be hooked up in the postamble. Therefore the tip and the tail of the gluon that connects the two particles has got to be ascribed two different colours, that of the previous and the new particle.

Another important particle is the *neutron*. It consists of three quarks like the proton, but it should have a forth particle to complete the set of two positive and two negative charges to make it neutral. If we try to set up all the internal connections between the three particles, we find that it is impossible. Only if we add a positive charge into the model we get a complete set of connections (see Fig. 7.4, page 153). That is consistent with Quantum chromodynamics (QCD) (see Footnote on page 154). The positive charge surrounded by a corresponding negative charge can probably best be characterized as a virtual positron. The force that keeps it in place has got to be related to *the weak force*, which is one of the basic forces of nature. As a conclusion it seems most probable that some kind of string theory has got to be applied in order to fully describe how matter is being built up.

Finally, links between quarks need not only be between quarks in the same hadron. The gluons may occasionally link up to a quark in another hadron, and it is those stray connections that have got to account for the forces tying baryons together into a nucleus. In Chapter 7 I'll try to get a more detailed picture of how these particles may be built-up.

5.10   The material body

In this model of space and matter a material body is built up of
nodes oscillating around axes with the same spatial direction. If it
were not so, the displacement fields from nearby nodes could not
cancel each other out due to destructive interference outside the
body and the energy would grow towards infinity. Also if the axes
of the nodes get into slightly different directions, it will result in
an increased field energy throughout space and thus force the axes
back into one common direction. This makes such a system very
stable.

Since all the displacements are along concentric circles around
the axes, all the resulting components have got to be in planes normal to these axes. If we organize a coordinate system with the z-axis
in the same direction as the rotational axes, the z-components becomes zero and deformations can be described by the xy-components.
This fact makes it possible to describe all displacements as the real
and imaginary parts of a complex number.

A material body consists of a great number of real and virtual
particles, which all are strings of oscillating nodes and all oscillates
with their own frequencies. Not two of the strings are allowed to
have coinciding peak values, so the peak values of all the strings
have got to fall in between each other. Hence the peaks occur with
a frequency that is like the sum of all the single strings. Each string
is supposed to have an energy given by a fixed constant, h, times
the frequency, f_p, where h is Plank's constant. Thus the material
body has an energy given by h times this frequency, $E = hf_b$, and
the frequency is given by $f_b = E/h$.

The inner frequency is not detectable when the body is at rest,
but when the body is in motion, the situation is different. Think
of the inner oscillation as a simple harmonic oscillation, and let the
the whole system move along with a group velocity, say v_g. This is
the velocity of the body.

To get an idea of what is going on, we can think of an over-simplified model of a material body. Say that a scalar wave moves along a circle with a speed c, and let the wavelength be like the circumference of the circle. Then the wave crest of the oscillation will be at the same point on the circle at every passage, e.g. at the top. Now let the circle move in some direction with a velocity v. Then in the course of one revolution the wave has passed from one wave crest to the next while the circle has moved a short distance. As seen from the outside, the passage from one wave crest through the trough and back to the next crest is seen as a wavelength of a wave moving with a speed, v.

There is, however, another wave that moves through the wave packet, which represents the body, with a speed far greater than c. To see this wave, one has got to make a more subtler mathematical approach. It will be necessary to first show that the mathematics that describes the deformations, is invariant by Lorentz transformations, which I have done in Chap. 4. Next let us think of a material body as an oscillatory system with amplitudes given by a function of position and time. If such a system moves along in space, it will behave like two waves, one moving with a group speed like the speed of the body, and one moving through the wave packet with a far higher speed. We can get an impression of these waves by first considering a body at rest, and then as seen from a Lorenz coordinate system moving with some speed, v (see Fig. 5.7).

By a similar, but more thorough consideration, it is possible to develop the Schrödinger equation (see Sec. 7.8):

$$\frac{-\hbar^2}{2m}\nabla^2\psi + V\psi = i\hbar\frac{\partial\psi}{\partial t}.$$

Even this equation is not complete because it only is developed for bodies moving with velocities far slower than the speed of light, and moreover, without taking into account the spin properties of material bodies. Hence a still more thorough examination is needed in order to get at the Dirac equation, which incorporates all known properties of matter except gravitation.

Consider a stationary oscillation system in one frame, (x, y, z):

$$\psi = A(x, y, z, t)e^{-i\omega t},$$

and boost it to another system that is moving with a speed, $v \ll c$, along the x-axis by setting:

$$
\begin{aligned}
t' &= \gamma(t - vx/c^2) \approx \gamma t, \\
x' &= \gamma(x - vt), \; y' = y, \; z' = z, \\
\gamma &= \frac{1}{\sqrt{(1 - v^2/c^2)}} \approx 1,
\end{aligned}
$$

In the moving frame the system represents two wave movements: One with group velocity, v_g, and one moving through the vave packet with a velocity, v_m:

$$
\begin{aligned}
\psi' &\approx A(x - vt, y, z, t)e^{-i\omega(t - vx/c^2)} \\
&\approx A(x - vt, y, z, t)\exp[i\omega \cdot v/c^2(x - c^2/v \cdot t)].
\end{aligned}
$$

and boost it to another system that is moving with a speed, $v \ll c$, along the x-axis by setting:

$$
\begin{aligned}
v_g &= v, \\
v_m &= c^2/v.
\end{aligned}
$$

The last equation represents the matter wave proposed by Louis de Broglie in 1924.

Figure 5.7: De Broglie's matter wave.

5.11 Summing-up

The idea that matter basically is disturbances in the spatial continuum is based on two kinds of oscillating nodes. Those witch oscillate between compression and rarefaction, and those witch oscillate between rotation in the right and left direction. Spherical waves go into the nodes, get reflected and move on to another node in an eternal round-dance. They obtain their lowest energy if they organize along chains of nodes separated by half a wavelength. Since all other constellations have higher energy, such chains of nodes are extremely stable.

Such strings of nodes do not displace any amount of spatial mass, or impose any resulting spin into the continuum because the nodes pairwise cancel their effects. An elementary material particle typically builds up a net displacement or rotation that move from node to node along the chain. The amplitude of the oscillation in the preamble gradually increases until it culminates in a highly exited node, say a "bubble", whereupon the oscillation fades away towards zero in the postamble. The bubble itself is filled up with spatial mass from the next node in the chain, and in this way the bubble moves on from node to node along the chain. The chain may either be a straight line in which case the particle moves along with the typical wave speed, or it may be curled up in which case the particle can move along with any speed lower than that.

Consider a straight string of right/left spinning nodes building up to an exited node. The nodes in the preamble will trigger a plane S-wave that gradually builds up to a maximum, and is expected to be caught up in the postamble where it fades away. However, a close examination reveals that the oscillation in the postamble is half a period out of phase with the oscillation in the preamble. This discrepancy is supposed to be corrected by a flip over of rotation to the opposite direction in the exited node. Photons are thought to be particles of this type. They move in straight lines, are polarized, possess spin, and are at the same time particles and waves.

An electric field is the negative of a velocity field that is maintained by virtual photons. A virtual photon is a short-lived photon that either brings with it a surplus or missing amount of spatial matter dubbed up or down polarized virtual photons respectively. Up and down polarized virtual photons are created by pair production and move in opposite directions for a short distance until they annihilate with other virtual particles in the environment. A moving up polarized photon drags spatial mass in the same direction as its own movement, and a down polarized photon pushes spatial mass in the opposite direction. In this way the velocity field is maintained.

An electron is a curled up string of nodes that emits down polarized virtual photons and spews out spatial mass like a source, and a positron is a corresponding particle that sucks in spatial mass like a sink and emits up polarized virtual particles. They are the engines that set up and maintain the electric field between them.

While the electric fields between leptons is a velocity field that is kept up by an exchange of free virtual particles, the forces between quarks are conveyed by unbroken strings along which the exchange of spatial mass is transported. It is the strength of the strings that represent the strong forces of nature.

Finally, since matter consists of oscillating nodes with all axes pointing in the same spatial direction, it is in principle possible to describe any displacement within a material body by only two components normal to that direction. Schröedinger took advantage of this property and developed his famous equation whereby the state of a system of material particles could be described by only one complex number.

Chapter 6

The oscillating node

Central in the idea of matter as disturbance energy in the spatial continuum, is the oscillating node. A single oscillating node, either if it is oscillating between compression and rarefaction, or oscillating between a right and left rotation, will quickly radiate away its energy and die out. However, if one could find systems of oscillating nodes where the radiation from one node finds its way into another node from which it is reflected and so on, the situation might be different. Perhaps there are constellations of nodes that keep the energy bound to the system. Then such a system could be identified as matter. In this chapter I'll study the character of such nodes and the interaction between them. The displacement fields around the nodes may either interfere in a constructive or a destructive manner, and if the interference is mostly destructive at some distance from the collection of nodes, the escape of energy might be dampened or even stopped. Then the energy would be confined and we would have a material body.

6.1 Elastodynamic waves

Waves in an elastic continuum are described by the Navier-Cauchy equation:

$$(\lambda_s + 2\mu_s)\nabla \text{div}\, \mathbf{u} - \mu_s \text{curl}\,\text{curl}\,\mathbf{u} + \mathbf{b} = \rho_s \ddot{\mathbf{u}}. \qquad (6.1)$$

where **u** is a small displacement, **b** is an outer force, μ_s and λ_s, are
Lamé's elastic constants, and ρ_s is the mass density of the spatial
continuum. By defining two new constants

$$c_g = \sqrt{\frac{\lambda_s + 2\mu_s}{\rho_s}}, \quad c_l = \sqrt{\frac{\mu_s}{\rho_s}}, \tag{6.2}$$

the N-C equation takes the form:

$$c_g{}^2 \nabla \operatorname{div} \mathbf{u} - c_l{}^2 \operatorname{curl} \operatorname{curl} \mathbf{u} + \frac{\mathbf{b}}{\rho_s} = \ddot{\mathbf{u}}.$$

According to *Helmholtz's decomposition theorem*, any vector field
that satisfies the condition:

$$[\nabla \times \mathbf{a}]_\infty \;=\; 0,$$

$$[\nabla \cdot \mathbf{a}]_\infty \;=\; 0,$$

may be written as the sum of an irrotational and a solenoidal part:

$$\mathbf{a} = -\nabla\phi + \nabla \times \mathbf{A}.$$

As the spatial continuum is of infinite extension, or nearly so, any
deformations have got to be confined to a finite part of space, so
this theorem will be applicable on all deformations. Hence the displacement field can be decomposed into the two properties:

$$\mathbf{u} = \hat{\mathbf{u}} + \tilde{\mathbf{u}},$$

where

$$\begin{aligned} \hat{\mathbf{u}} &= -\nabla\phi = -\operatorname{grad}\phi, \\ \tilde{\mathbf{u}} &= \nabla \times \psi = \operatorname{curl}\psi, \quad \operatorname{div}\psi = 0, \end{aligned}$$

and if an outer force is applied, then **b** is supposed to decompose
in the same way.

By the mathematical identities, $\operatorname{curl}\operatorname{grad}(\bullet) = 0$ and $\operatorname{div}\operatorname{curl}(\bullet) = 0$, (the curl of a gradient and the divergence of a curl is always like zero) the Navier-Cauchy equation can be divided into two independent equations, one for an irrotational field

$$\operatorname{grad} \operatorname{div} \hat{\mathbf{u}} = \frac{1}{c_g^2}\ddot{\hat{\mathbf{u}}} - \frac{\hat{\mathbf{b}}}{\lambda_s + 2\mu_s}, \tag{6.3}$$

and the other for a solenoidal field:

$$-\operatorname{curl}\operatorname{curl}\tilde{\mathbf{u}} = \frac{1}{c_l^2}\ddot{\tilde{\mathbf{u}}} - \frac{\tilde{\mathbf{b}}}{\mu_s} \tag{6.4}$$

Operating on the two equations with the $\nabla\cdot$ and $\nabla\times$ operator respectively, and setting $\mathbf{b}$ to zero, we obtain:

$$\nabla^2(\operatorname{div}\mathbf{u}) - \frac{1}{c_g^2}\frac{\partial^2(\operatorname{div}\mathbf{u})}{\partial t^2} = 0,$$

$$\nabla^2(\operatorname{curl}\mathbf{u}) - \frac{1}{c_l^2}\frac{\partial^2(\operatorname{curl}\mathbf{u})}{\partial t^2} = 0.$$

These are two wave equations where the dilatation, div $\mathbf{u}$, satisfies a wave moving with the speed c_g, while the rotational component curl $\mathbf{u}$, satisfies a wave moving with the speed c_l. In *fact the Propagation theorem for isotropic bodies* states that if a body is isotropic, then a wave is either longitudinal, in which case $c = c_g$, or transversal, in which case $c = c_l$ [6, p. 256].

Both waveforms can either be plane waves or central symmetrical, shell-formed, waves. A plane longitudinal wave propagates in accordance with the formula:

$$\operatorname{div}\mathbf{u} = f(c_g t - x) + g(c_g t + x), \tag{6.5}$$

and a shell-formed wave is described by:

$$\operatorname{div}\mathbf{u} = \frac{f(c_g t - r)}{r} + \frac{g(c_g t + r)}{r}. \tag{6.6}$$

Plane waves can never be confined. They will move in certain directions forever and get lost somewhere at the border of the universe. However, spherical waves can both move outwards from a center or towards it, so what happens to a converging spherical wave when it reaches the focal point? It certainly cannot disappear into that point, but has got to be reflected or reappear in some obscure form. The last alternative will probably be forbidden for symmetry reasons, if both the waves and the spatial continuum are perfect enough.

The splitting of the Navier-Cauchy equation into one irrotational and one solenoidal part, allows us to examine these two parts separately and thereby simplify the strain-stress relation immensely by reducing the elastic constants to only one single constant (the wave speed) in each equation ($c_g \neq c_l$). We see from 6.2 that the two wave speeds are related to each other with a fixed constant given by the relation $c_g = \sqrt{2 + \lambda_s/\mu_s} \cdot c_l$, where c_g might be about the double of c_l.

The energy in a deformation field with no surface trajectories is given by *Kelvin's theorem* [6, p. 208]:

$$E = \int_B \left[\frac{1}{2} \rho_s \dot{\mathbf{u}}^2 + \frac{1}{2} (\lambda_s + 2\mu)(\operatorname{div} \mathbf{u})^2 + \frac{1}{2} \mu(\operatorname{curl} \mathbf{u})^2 \right] dv.$$

The theorem states that to a curl $\mathbf{u}$, a div $\mathbf{u}$, and a velocity field $\dot{\mathbf{u}}$ there always corresponds an energy equal to E, but it does not tell exactly where in the field the energy is to be found. That is because the equation is only correct if the three elements are integrated all over deformed space, but then there is no additional energy around. With this restriction in mind, the local energy density, e, in a spatial continuum of infinite extension can all the same be defined as[1]:

$$e = \frac{1}{2} \rho_s \dot{\mathbf{u}}^2 + \frac{1}{2} (\lambda_s + 2\mu_s)(\operatorname{div} \mathbf{u})^2 + \frac{1}{2} \mu_s(\operatorname{curl} \mathbf{u})^2. \qquad (6.7)$$

because the relation between deformation and energy always is given by this formula.

---

[1] The corresponding expression for the energy in an electromagnetic field has the same limitation, but nonetheless it is usually interpreted as the local energy density.

6.2 Irrotational standing waves

Generally, standing waves may be seen as a result of interference between two waves with the same amplitude travelling in opposite directions. Take as an example a vibrating string. It can be treated mathematically as progressive waves travelling back and forth between the two fixed endpoints of the string where they are reflected. The wavelength is found by solving the wave equation for the string, and setting the amplitudes at the endpoints to zero. We find that wavelengths of 2, 1, 2/3, 1/2, . . . times the length of the string are possible. When we pluck a guitar string, however, we pull it out in one point such that the string becomes like a broken line before we release it. This is much different from a sinusoid shape, so the string does not oscillate with just one of these harmonics. To get a correct picture of the oscillation, we have got to represent the oscillation as a Fourier series. What we get is a superposition of a series of allowable frequencies with amplitudes a_0, a_1, a_2, $\cdots$, and so forth.

To see if standing waves also can occur in the spatial continuum, it will be necessary to first solve the wave equation for the set-up we want to examine with respect on the displacement vector, **u**, and then chose the integration constants such that **u** becomes zero at some fixed endpoints.

The first case I want to study is if a singularity in the form of an oscillating node can be an endpoint in an irrotational standing wave. To achieve that, I shall imagine a thought experiment with a rigid shell in the spatial continuum and solve the wave equation for concentric compression waves inside the shell. First we write the Navier-Cauchy equation for an irrotational field with no external forces 3.9:

$$\text{grad}\, div\mathbf{u} = \frac{1}{c_g^2}\ddot{\mathbf{u}},$$

where c_g is the speed of longitudinal waves, and then consider a central symmetric deformation where $\mathbf{u} = f(r)\hat{\mathbf{r}}$. $\hat{\mathbf{r}}$ is a unit vector in the direction of **r**. From the identities:

$$\mathrm{div}\, f\hat{\mathbf{r}} = 2\frac{f}{r} + f',$$

and

$$\mathrm{grad}\, f = f'\hat{\mathbf{r}},$$

we obtain the identity:

$$\mathrm{grad}\,\mathrm{div}\, f = \left(f'' + \frac{2f'}{r} - \frac{2f}{r^2} \right)\hat{\mathbf{r}}.$$

By these identities, the N-C equation takes the form:

$$\left(u'' + 2\frac{u'}{r} - 2\frac{u}{r^2} \right)\hat{\mathbf{r}} \;=\; \frac{\ddot{u}}{c_g^2}\hat{\mathbf{r}}. \qquad (6.8)$$

It can be divided into two equations by the product method, one of which can be integrated directly, and the other solved by the Frobenius' method. I have put the solving at the end of the section on page 114 in case someone wants to check the solution, but here I only write the solution directly:

$$
\begin{aligned}
\mathbf{u}(\mathbf{r}, t) \;=\;& [C_0 \sin(c_1 pt) - C_2 \cos(c_1 pt)] \\
& \times \left\{ \frac{1}{r}\left[A_0 p \sin(pr) - \frac{3A_3}{p^2}\cos(pr) \right] \right.\\
& \left. + \frac{1}{r^2}\left[\frac{3A_3}{p^3}\sin(pr) + A_0 \cos(pr) \right] \right\}\hat{\mathbf{r}}. \qquad (6.9)
\end{aligned}
$$

This is the complete solution for concentric waves in the spatial continuum. In order to see if there may be a standing wave in the space between the rigid sphere and the center, **u** has got to be zero at these two positions. The time dependent term above represents a simple harmonic oscillator, and the constants C_0 and C_1 only describe different phase shifts, so for convenience I set C_0 to zero. By drawing a graph of the function above with different values of A_0, we see from Figure 6.1 that it is only when A_0 is zero that we

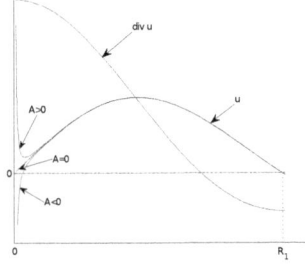

Figure 6.1: Shell waves.
Standing wave between the inside of a rigid shell with radius R_1 and the center node.

achieve the correct border condition at the center node. Hence the equation above reduces to

$$\mathbf{u}(\mathbf{r}, t) = A \cos(c_g p t) \cdot \left[\frac{\sin(pr)}{p^3 r^2} - \frac{\cos(pr)}{p^2 r} \right] \hat{\mathbf{r}}, \qquad (6.10)$$

where $A = -3A_3 C_1$. The condition that $\mathbf{u} = 0$ at the radius of the rigid sphere, R_n, is met when:

$$\frac{\sin(pR_n)}{p^3 R_n{}^2} - \frac{\cos(pR_n)}{p^2 R_n} = 0,$$
$$\tan(pR_n) = pR_n.$$

There is a solution for every $\tan(pR_n) = pR_n$, and we also find that the first radius to be considered is when $R_1 \approx 9/2p$ so the standing wave can have several wavelengths inside the shell.

Like with the guitar string, it is difficult to start the oscillation with a smooth oscillation as described above. In stead it is more likely that given a certain cavity, we can stir the inside in some irregular ways. The result will be a series of allowed oscillations with amplitudes a_0, a_1, a_2, $\cdots$.

In order to check that $\mathbf{u}$ really approaches zero when $r \to 0$, we can apply l'Hôpital's rule:

$$\lim_{r \to 0} \frac{\sin(pr) - pr \cos(pr)}{(pr)^2} = \lim_{r \to 0} \frac{[\sin(pr) - pr \cos(pr)]'}{[(pr)^2]'}$$
$$= \lim_{r \to 0} \frac{1}{2} \sin(pr)$$
$$= 0.$$

By applying the identity, div $(f(r)\hat{\mathbf{r}}) = 2f(r)/r + f'(r)$, on 6.10, we can find how div $\mathbf{u}$ varies as a function of $\mathbf{r}$:

$$\text{div } \mathbf{u} = \frac{2u}{r} + \frac{\partial u}{\partial r}$$

$$= A\cos(c_g pt)\left[\frac{2\sin(pr)}{p^3 r^3} - \frac{2\cos(pr)}{p^2 r^2}\right.$$

$$\left. -\frac{2\sin(pr)}{p^3 r^3} + \frac{\cos(pr)}{p^2 r^2} + \frac{\cos(pr)}{p^2 r^2} + \frac{p\sin(pr)}{p^2 r}\right].$$

Hence:

$$\text{div } \mathbf{u} = A\cos(c_g pt)\frac{\sin(pr)}{pr}. \tag{6.11}$$

We also have that[2]:

$$\lim_{r\to 0}(\text{div } \mathbf{u}) = A\cos(c_g pt).$$

It is noteworthy that even if the node itself is a mathematical singularity such that $r \neq 0$, the limit of both the displacement $\mathbf{u}$ and the divergence of $\mathbf{u}$ have finite limits when r approaches zero.

This thought experiment shows us that standing waves may be established between a single node and a firm concentric shell, but it also indicates that individual points in space may be nodes in a more complex system of waves that are moving between singularities were they are reflected. The reflection is without a phase shift: An incoming wave peak is reflected as a wave peak.

Solving Equation 6.8

The equation to be solved is:

$$\left(u'' + 2\frac{u'}{r} - 2\frac{u}{r^2}\right)\hat{\mathbf{r}} = \frac{\ddot{u}}{c_g^2}\hat{\mathbf{r}}.$$

---

[2]Note that $\lim_{x\to 0} \frac{\sin(x)}{x} = 1$. By the way, the property, $\sin(x)/x$, occurs so often in physics that it has got its own name, the *sinc function*, or the *normalized sinc function*, $\text{sinc}(x) = \sin(\pi x)/\pi x$. It will come to play an important role when I shall try to figure out a model of a material particle.

It can be solved by the product method by setting.

$$\mathbf{u}(\mathbf{r}, t) = F(r) \cdot G(t)\hat{\mathbf{r}} \tag{6.12}$$

We obtain:

$$F''G + \frac{2}{r}F'G - \frac{2}{r^2}FG = \frac{1}{c_g{}^2}F\ddot{G},$$

$$\frac{F''}{F} + \frac{2F'}{rF} - \frac{2}{r^2} = \frac{\ddot{G}}{c_g{}^2 G}$$

Since the left and right side of this equation only depend of r and t respectively, they can both be set to the same constant, say $-p^2$, and hence they can be separated into two separate, ordered equations:

$$F'' + \frac{2}{r}F' + \frac{-2 + p^2 r^2}{r^2}F = 0, \tag{6.13}$$

$$\ddot{G} + c_g{}^2 p^2 G = 0. \tag{6.14}$$

The second equation has the general solution:

$$G(t) = C_0 \sin[c_g p t] - C_1 \cos(c_g p t), \tag{6.15}$$

while the first equation may be solved by the FROBENIUS method (see e.g. [10, Sec. 4.4]: *Any differential equation of the form*

$$y'' + \frac{b(x)}{x}y' + \frac{c(x)}{x^2}y = 0$$

where the functions b(x) and c(x) are analytic at x=0, has at least one solution that can be represented in the form

$$y(x) = x^r (a_0 + a_1 x + a_2 x^2 + \cdots) \tag{6.16}$$

where the exponent r may be any (real or complex) number (and r is chosen so that $a_0 \neq 0$).

The equation also has a second solution (such that these two so-lutions are linearly independent) that may be similar to 6.16 (with a different r and different coefficients) or may contain a logarithmic term.)

Temporary replacing r with x and F with y in Equation 6.13 yields

$$y'' + \frac{2}{x}y' + \frac{-2 + p^2 x^2}{x^2}y = 0.$$

This equation has at least one solution of the form

$$y(x) = x^r \sum_{m=0}^{\infty} a_m x^m = \sum_{m=0}^{\infty} a_m r^{m+r}.$$

The first and second derivative becomes

$$y'(x) = \sum_{m=0}^{\infty} a_m(m+r)x^{m+r-1},$$

$$y''(x) = \sum_{m=0}^{\infty} a_m(m+r)(m+r-1)x^{m+r-2}.$$

We expand $b(x)$ and $c(x)$ in power series with coefficients

$$b_0 = 2, \ b_{1,2,\dots} = 0, \ c_0 = -2, \ c_1 = 0, \ c_2 = p^2, \ c_{3,4,\dots} = 0.$$

The *indicial equation* $r(r-1) + b_0 r + c_0 = 0$ takes the form

$$r(r-1) + 2r - 2 = 0$$
$$r_1 = 1, \quad r_2 = -2. \tag{6.17}$$

Here we have a *case 3* situation (roots differing by an integer, $r_1 - r_2 = 3$). We also note that $r_1 - r_2 > 0$ so we have the two solutions

$$
\begin{aligned}
y_1(x) &= x^{r_1}(a_0 + a_1 x + a_2 x^2 + \cdots), & (6.18) \\
y_2(x) &= k y_1(x)\ln x + x^{r_2}(A_0 + A_1 x + A_2 x^2 + \cdots), & (6.19)
\end{aligned}
$$

where k may or may not be equal to zero.

The solution and derivatives inserted into the original equation

$$x^2 \sum_{m=0}^{\infty} a_m(m+r)(m+r-1)x^{m+r-2}$$

$$+2x \sum_{m=0}^{\infty} a_m(m+r)x^{m+r-1}$$

$$-2 \sum_{m=0}^{\infty} a_m x^{m+r} + p^2 x^2 \sum_{m=0}^{\infty} a_m x^{m+r} = 0.$$

From the indicial Equation 6.17 we choose $r=1$ and collect the parts with the same power of x

$$a_s(s+1)sx^{s+1} + 2a_s(s+1)x^{s+1}$$
$$-2a_s x^{s+1} + p^2 a_{s-2} x^{s+1} = 0,$$
$$a_s(s^2 + s + 2s + 2 - 2) + a_{s-2}p^2 = 0,$$
$$a_s \cdot s(s+3) = -a_{s-2}p^2,$$
$$a_s = a_{s-2}\frac{-p^2}{s(s+3)},$$

and develop the even coefficients of a_n

$$a_0 \cdot 0(0+3) = 0,$$
$$a_0 = a,$$
$$a_2 = a_0\frac{(-p^2)}{2 \cdot 5} = a\frac{(-p^2)}{2 \cdot 5},$$
$$a_4 = a_2\frac{(-p^2)}{4 \cdot 7} = a\frac{(-p^2)^2}{2 \cdot 5 \cdot 4 \cdot 7},$$
$$a_6 = a_4\frac{(-p^2)}{6 \cdot 9} = a\frac{(-p^2)^3}{2 \cdot 5 \cdot 4 \cdot 7 \cdot 6 \cdot 9},$$
$$\cdots \qquad \cdots\cdots\cdots\cdots\cdots$$
$$a_m = a(-1)^{\frac{m}{2}} \cdot p^m \frac{3 \cdot (m+2)}{(m+3)!},$$
$$= a(-1)^{\frac{m}{2}} \cdot p^m \left[\frac{3}{(m+2)!} - \frac{3}{(m+3)!}\right], \quad m = 0,2,4,6,\cdots,$$
$$a_{2n} = a(-1)^n p^{2n} \left[\frac{3}{(2n+2)!} - \frac{3}{(2n+3)!}\right], \quad n = 0,1,2,\cdots.$$

The odd coefficients of a_n are all zero:

$$a_1 \cdot 1(1+3) = 0,$$
$$a_1 = 0,$$
$$a_3 = a_1 \frac{(-p^2)}{3 \cdot 6} = 0,$$
$$a_5 = a_7 = a_9 = \cdots = 0.$$

According to 6.18 the first solution is:

$$
\begin{aligned}
y_1 &= x^{r_1} \sum_{n=0}^{\infty} a_{2n} x^{2n} \\
&= x \sum_{n=0}^{\infty} a_0 (-1)^n p^{2n} \left[\frac{3}{(2n+2)!} - \frac{3}{(2n+3)!} \right] x^{2n} \\
&= 3a_0 \left(\frac{x}{2!} - \frac{p^2 x^3}{4!} + \frac{p^4 x^5}{6!} - \cdots - \frac{x}{3!} + \frac{p^2 x^3}{5!} - \frac{p^4 x^5}{7!} + \cdots \right) \\
&= \frac{3a_0}{p^2 x} \left[1 - \left(1 - \frac{(px)^2}{2!} + \frac{(px)^4}{4!} - \frac{(px)^6}{6!} + \cdots \right) \right] \\
&\quad \frac{3a_0}{p^3 x^2} \left[-px + \left(px - \frac{(px)^3}{3!} + \frac{(px)^5}{5!} - \frac{(px)^7}{7!} + \cdots \right) \right], \\
&= \frac{3a_0}{p^3 x^2} \sin(px) - \frac{3a_0}{p^2 x} \cos(px).
\end{aligned}
$$

There may be another solution 6.19 of the form

$$y_2 = k y_1 \ln(x) + (A_0 + A_1 x + A_2 x^2 + \cdots),$$

where k might or might not be zero. First I try the solution where k is set to zero. From the indicial Equation 6.17 we choose $r_2 = -2$ and collect the parts with the same power of x.

$$
\begin{aligned}
A_m(m-2)(m-3)x^{m-2} &+ 2A_m(m-2)x^{m-2} \\
-2A_m x^{m-2} &+ p^2 A_m x^m = 0, \\
A_s[(s-2)(s-3) + 2(s-2) - 2] &= -p^2 A_{s-2}, \\
A_s(s-3)s &= -p^2 A_{s-2}, \qquad\qquad (6.20) \\
A_s = A_{s-2} \frac{-p^2}{s(s-3)}, &\quad (s \neq 0, 3),
\end{aligned}
$$

and develop the coefficients of A_n

$$A_0(0-3)0 \;=\; 0,$$

$$A_0 \;=\; A_0,$$

$$A_2 \;=\; -A_0\frac{(-p^2)}{2\cdot 1},$$

$$A_4 \;=\; -A_2\frac{-p^2}{4\cdot 1} = A_0\frac{(-p^2)^2}{1\cdot 2\cdot 4},$$

$$A_6 \;=\; -A_4\frac{-p^2}{3\cdot 6} = A_0\frac{(-p^2)^3}{1\cdot 2\cdot 3\cdot 4\cdot 6},$$

$$\cdots\cdots$$

$$A_n \;=\; -A_0(-1)^{\frac{n}{2}}\frac{p^n(n-1)}{n!}, \qquad n = 0,2,4,\cdots,$$

$$A_n \;=\; A_0(-1)^{\frac{n}{2}}\left[\frac{p^n}{n!} - \frac{p^n}{(n-1)!}\right], \qquad n = 2,4,6,\cdots,$$

$$A_{2s} \;=\; A_0(-1)^s\left[\frac{p^{2s}}{(2s)!} - \frac{p^{2s}}{(2s-1)!}\right], \qquad s = 1,2,3,\cdots.$$

We see from 6.20 that $A_1 = 0$, but that A_3 may still be $\neq 0$. We get the odd coefficients of A_n:

$$A_1(1-3)1 \;=\; 0,$$

$$A_1 \;=\; 0,$$

$$A_3(3-3)3 \;=\; A_1(-p^2) = 0,$$

$$A_3 \;=\; A_3,$$

$$A_5 \;=\; A_3\frac{(-p^2)}{5\cdot 2},$$

$$A_7 \;=\; A_5\frac{(-p^2)}{7\cdot 4} = A_3\frac{(-p^2)^2}{2\cdot 4\cdot 5\cdot 7},$$

$$\cdots\cdots$$

$$A_n \;=\; A_3(-1)^{\frac{n-3}{2}}\cdot\frac{p^{n-3}\cdot 3\cdot(n-1)}{n!}, \qquad n = 3,5,7,\cdots,$$

$$A_{2s+3} \;=\; 3A_3(-1)^s\left[\frac{p^{2s}}{(2s+2)!} - \frac{p^{2s}}{(2s+3)!}\right], \qquad s = 0,1,2,\cdots.$$

We first seek the partial solution of 6.19 with the *odd* coefficients of A_n set to zero:

$$y_{2_1} = x^{-2}\left[A_0 + \sum_{s=1}^{\infty} A_0(-1)^s\left(\frac{p^{2s}}{(2s)!} - \frac{p^{2s}}{(2s-1)!}\right)x^{2s}\right]$$

$$= A_0x^{-2}\left(1 - \frac{(px)^2}{2!} + \frac{(px)^4}{4!} - \frac{(px)^6}{6!} + \cdots\right)$$

$$+ A_0px^{-1}\left(\frac{(px)^1}{1!} - \frac{(px)^3}{3!} + \frac{(px)^5}{5!} - \cdots\right),$$

$$y_{2_1} = \frac{A_0p}{x}\sin(px) + \frac{A_0}{x^2}\cos(px).$$

Next we seek the partial solution of 6.19 with the *even* coefficients of A_n set to zero:

$$y_{2_2} = x^{-2}\sum_{s=0}^{\infty} A_{2s+3}x^{2s+3}$$

$$= x^{-2}\sum_{s=0}^{\infty} 3A_3(-1)^sp^{2s}\left[\frac{1}{(2s+2)!} - \frac{1}{(2s+3)!}\right]x^{2s+3}$$

$$= 3A_3\left(\frac{x}{2!} - \frac{p^2x^3}{4!} + \frac{p^4x^5}{6!} - \cdots - \frac{x}{3!} + \frac{p^2x^3}{5!} - \frac{p^4x^5}{7!} + \cdots\right)$$

$$= \frac{3A_3}{p^2x}\left[1 - \left(1 - \frac{(px)^2}{2!} + \frac{(px)^4}{4!} - \frac{(px)^6}{6!} + \cdots\right)\right]$$

$$\frac{3A_3}{p^3x^2}\left[-px + \left(px - \frac{(px)^3}{3!} + \frac{(px)^5}{5!} - \frac{(px)^7}{7!} + \cdots\right)\right],$$

$$y_{2_2} = \frac{3A_3}{p^3x^2}\sin(px) - \frac{3A_3}{p^2x}\cos(px).$$

We notice that the partial solution y_{2_2} is identical to the partial solution y_1, hence the complete solution has got to be of the form

$$y = ky_1\ln x + y_{2_1} + y_{2_2}.$$

We can test this equation in order to see if $k \neq 0$.

$$y = [y_1] \cdot k\ln x + [y_{2_1} + y_{2_2}]$$

$$y' = [y_1'] \cdot k\ln x + y_1\frac{1}{x} + [y_{2_1}' + y_{2_2}']$$

$$y'' = [y_1''] \cdot k\ln x + \frac{2y_1'}{x} - y_1\frac{1}{x^2} + [y_{2_1}'' + y_{2_2}''].$$

We already know that the expressions in square brackets are partial solutions of 6.13 and zeroes out, hence it only remains to see if the rest of the terms may be another partial solution with $k \neq 0$ or if k has got to be zero. We have

$$x^2 \left(y_1' \frac{2}{x} - y_1 \frac{1}{x^2} \right) + 2x \left(y_1 \frac{1}{x} \right)$$

$$= 2xy_1' - y_1 + 2y_1$$

$$= 2xy_1' + y_1$$

$$= 2x \left[-\frac{2a}{px^3} \sin(px) + \frac{2a}{x^2} \cos(px) + \frac{ap}{x} \sin(px) \right]$$

$$+ \frac{a}{px^2} \sin(px) - \frac{a}{x} \cos(px)$$

$$= \frac{3a}{x} \cos(px) - \frac{3a}{px^2} \sin(px) + 2ap \sin(px).$$

This expression cannot be zero for any range of r, so k has got to be zero, and the solution of the differential equation is given by

$$y = \frac{1}{x} \left[A_0 p \sin(px) - \frac{3A_3}{p^2} \cos(px) \right]$$

$$+ \frac{1}{x^2} \left[\frac{3A_3}{p^3} \sin(px) + A_0 \cos(px) \right],$$

and by reentering $F = y$ and $r = x$ we finally acquire the solution of 6.13:

$$F = \frac{1r \left[A_0 p \sin(pr) - \frac{3A_3}{p}^2 \right]}{\cos} (pr) \right] + \frac{1}{r^2} \left[\frac{3A_3}{p^3} \sin(pr) + A_0 \cos(pr) \right]$$

$$(6.21)$$

By 6.12, 6.15, and 6.21 the displacement vector becomes:

$$\mathbf{u}(\mathbf{r}, t) = [C_0 \sin[c_g pt] - C_1 \cos(c_g pt)]$$

$$\cdot \left\{ \frac{1}{r} \left[A_0 p \sin(pr) - \frac{3A_3}{p^2} \cos(pr) \right] \right.$$

$$\left. + \frac{1}{r^2} \left[\frac{3A_3}{p^3} \sin(pr) + A_0 \cos(pr) \right] \right\}, \qquad (6.22)$$

which is the solution of Equation 6.8.

6.3   Solenoidal standing waves

The Navier-Cauchy equation for solenoidal deformations is 3.10:

$$\operatorname{curl}\operatorname{curl}\mathbf{u} = -\frac{1}{c^2}\ddot{\mathbf{u}}.$$

Solving this equation for a deformation around a singularity may be a bit tricky, so first I will try to guess a solution of the form (The *hat* sign over $\mathbf{r}$ and $\mathbf{m}$ indicates that they are unit vectorsFru):

$$\mathbf{u}(\mathbf{r}.t) = g(r,t)(\hat{\mathbf{m}}\times\hat{\mathbf{r}}),$$

where $\hat{\mathbf{m}}$ is a fixed direction in space[3]. Then the equation we have got to solve becomes:

$$\operatorname{curl}\operatorname{curl}\left[g(r,t)(\hat{\mathbf{m}}\times\hat{\mathbf{r}})\right] = \frac{1}{c_l^2}\ddot{g}(r,t)(\hat{\mathbf{m}}\times\hat{\mathbf{r}}) \tag{6.23}$$

It has the solution (see end of section page 124):

$$\mathbf{u} = \cos(c_l kt)\left[\frac{\sin(kr)}{(kr)^3} - \frac{\cos(kr)}{(kr)^2}\right](\mathbf{r}\times\mathbf{M}), \tag{6.24}$$

By some of the identities:

$$\begin{aligned}
\operatorname{curl}(\phi\mathbf{A}) &= \phi\operatorname{curl}(\mathbf{A}) + \operatorname{grad}\phi\times\mathbf{A}, \\
\operatorname{curl}(\mathbf{c}\times\mathbf{r}) &= 2\mathbf{c}, \\
\operatorname{curl}\mathbf{r} &= 0, \\
\operatorname{grad}(\mathbf{c}\times\mathbf{r}) &= \mathbf{c}, \\
\mathbf{A}\times(\mathbf{B}\times\mathbf{C}) &= (\mathbf{A}\mathbf{C})\mathbf{B} - (\mathbf{A}\mathbf{B})\mathbf{C},
\end{aligned}$$

where $\mathbf{c}$ is a constant vector, $\mathbf{r}$ a radius vector, ϕ a scalar, and a little work, we can find $\operatorname{curl}\mathbf{u}$. With $\xi(t) = M\cos(c_l kt)$ we obtain:

$$\operatorname{curl}\mathbf{u} = \xi(t)\frac{3\sin(kr) - 3kr\cos(kr) - k^2 r^2\sin(kr)}{k^3 r^3}$$
$$\cdot[\hat{\mathbf{m}} - (\hat{\mathbf{m}}\times\hat{\mathbf{r}})]\hat{\mathbf{r}}$$

---

[3]The probelem is much the same as finding the vector potential around a magnetic dipole, which is:
$\mathbf{A}(\mathbf{r}) = \mu_0/4\pi r^3 \cdot (\mathbf{m}\times\mathbf{r})$.

It might be of interest to see how curl **u** behaves when r approaches zero. By applying l'Hôpital's rule we find:

$$\lim_{r \to 0} \operatorname{curl} \mathbf{u} = M \cos(c_l k t). \tag{6.25}$$

This interesting result confirms that curl **u** has a finite value in the vicinity of an oscillating singularity. Another interesting conclusion from the results above is that all displacement components are normal to the fixed direction of **M**, and the radius vector **r**. Say that we organize the coordinate axis in a Cartesian coordinate system such that **M** is in the z-direction. Then the displacement field lines form concentric circles around the z-axis, and absolutely all components of **u** are in the xy-plane. This makes it possible to express **u** as components of a complex number $\xi = \xi(\mathbf{r}, t)$ where ξ is a complex function of the position vector, **r**, and the time, t. Provided that all the nodes in a system have the same rotational direction, then the entire solenoidal field can be expressed as a simple sum of all the complex components generated by all the nodes in the system.

The wave equation:

$$\nabla^2 \mathbf{u} = \frac{1}{c^2} \ddot{\mathbf{u}}.$$

may be written as a complex function:

$$\nabla^2 \xi = \frac{1}{c^2} \ddot{\xi}.$$

But perhaps an even more useful equation can be obtained by defining a new property $\psi = \dot{\xi}$, which becomes a complex representation of the velocity vector field $\dot{\mathbf{u}}$ [$\psi = f(\mathbf{r}, t)$]. Then by taking the time derivative on both sides, the above equation takes the form

$$\nabla^2 \psi = \frac{1}{c^2} \ddot{\psi}. \tag{6.26}$$

The squared value of ψ is a real property that represents the kinetic field energy density by the equation:

$$e_{kin} = \frac{\rho_s}{2} |\psi \cdot \psi^*|.$$

In free oscillations the field energy divides equally between potential and kinetic energy, hence the field energy density can be expressed as:

$$e = \rho_s \psi^2.$$

I will return to this representation when I discuss the Schrödinger wave equation in another paper.

Solving Equation 6.23

The equation to be solved is:

$$\text{curl}\,\text{curl}\,[g(r,t)(\hat{\mathbf{m}} \times \hat{\mathbf{r}})] = \frac{1}{c_l^2}\ddot{g}(r,t)(\hat{\mathbf{m}} \times \hat{\mathbf{r}}), \qquad (6.27)$$

where $\hat{\mathbf{m}}$ is a constant vector. It can be solved by applying the identities:

$$
\begin{aligned}
\text{curl}\,(\phi \mathbf{A}) &= \phi\,\text{curl}\,\mathbf{A} + \text{grad}\,\phi \times \mathbf{A}, \\
\text{curl}\,(\mathbf{c} \times \mathbf{r}) &= 2\mathbf{c}, \\
\text{curl}\,\mathbf{r} &= 0, \\
\text{grad}\,(\mathbf{c} \cdot \mathbf{r}) &= \mathbf{c}, \\
\mathbf{A} \times (\mathbf{B} \times \mathbf{C}) &= (\mathbf{AC})\mathbf{B} - (\mathbf{AB})\mathbf{C},
\end{aligned}
$$

where $\mathbf{c}$ is a constant vector, $\mathbf{r}$ a radius vector, and ϕ a scalar field. We develop:

$$
\begin{aligned}
\text{curl}\,[g(\hat{\mathbf{m}} \times \hat{\mathbf{r}})] &= \text{curl}\left[\frac{g}{r}(\hat{\mathbf{m}} \times \mathbf{r})\right] \\
&= \frac{g}{r}\text{curl}\,(\hat{\mathbf{m}} \times \mathbf{r}) + \text{grad}\left(\frac{g}{r}\right) \times (\hat{\mathbf{m}} \times \mathbf{r}) \\
&= \frac{2g}{r}\hat{\mathbf{m}} + \left(\frac{g'}{r} - \frac{g}{r^2}\right)\hat{\mathbf{r}} \times (\hat{\mathbf{m}} \times \mathbf{r}), \\
&= \frac{2g}{r}\hat{\mathbf{m}} + \left(g' - \frac{g}{r}\right)[(\hat{\mathbf{r}} \cdot \hat{\mathbf{r}})\hat{\mathbf{m}} - (\hat{\mathbf{r}} \cdot \hat{\mathbf{m}})\hat{\mathbf{r}}], \\
&= \left(\frac{g}{r} + g'\right)\hat{\mathbf{m}} - \left(\frac{g'}{r} - \frac{g}{r^2}\right)(\mathbf{r} \cdot \hat{\mathbf{m}})\hat{\mathbf{r}},
\end{aligned}
$$

and further:

$$\operatorname{curl}\operatorname{curl}[g(\hat{\mathbf{m}} \times \hat{\mathbf{r}})]$$

$$= \left(\frac{g}{r} + g'\right)\operatorname{curl}\hat{\mathbf{m}} + \operatorname{grad}\left(\frac{g}{r} + g'\right) \times \hat{\mathbf{m}}$$

$$- \left(\frac{g'}{r} - \frac{g}{r^2}\right)(\mathbf{r} \cdot \hat{\mathbf{m}})\operatorname{curl}\hat{\mathbf{r}} - \operatorname{grad}\left[\left(\frac{g'}{r} - \frac{g}{r^2}\right)(\mathbf{r} \cdot \hat{\mathbf{m}})\right] \times \hat{\mathbf{r}}$$

$$= \left(\frac{g'}{r} - \frac{g}{r^2} + g''\right)(\hat{\mathbf{r}} \times \hat{\mathbf{m}}) - \left(\frac{g'}{r} - \frac{g}{r^2}\right)'\hat{\mathbf{r}}(\mathbf{r} \cdot \hat{\mathbf{m}}) \times \hat{\mathbf{r}}$$

$$- \left(\frac{g'}{r} - \frac{g}{r^2}\right)\operatorname{grad}(\mathbf{r} \cdot \hat{\mathbf{m}}) \times \hat{\mathbf{r}}$$

$$= \left(\frac{g'}{r} - \frac{g}{r^2} + g''\right)(\hat{\mathbf{r}} \times \hat{\mathbf{m}}) - \left(\frac{g'}{r} - \frac{g}{r^2}\right)(\hat{\mathbf{m}} \times \hat{\mathbf{r}})$$

$$= \left(g'' + \frac{2g'}{r} - \frac{2g}{r^2}\right)(\hat{\mathbf{r}} \times \hat{\mathbf{m}}).$$

With these terms inserted into 6.27 we obtain:

$$\left(g'' + \frac{2g'}{r} - \frac{2g}{r^2}\right)(\hat{\mathbf{r}} \times \hat{\mathbf{m}}) = \frac{1}{c_l^2}\ddot{g}(\hat{\mathbf{r}} \times \hat{\mathbf{m}}),$$

or the scalar differential equation:

$$\left(g'' + \frac{2g'}{r} - \frac{2g}{r^2}\right) = \frac{1}{c_l^2}\ddot{g},$$

which is exactly like 6.8 except for the value of c, and it yields a corresponding solution:

$$
\begin{aligned}
\mathbf{u} &= M\cos(c_l k t)\left[\frac{\sin(kr)}{k^3 r^2} - \frac{\cos(kr)}{k^2 r}\right](\hat{\mathbf{r}} \times \hat{\mathbf{m}}), \\
&= \cos(c_l k t)\left[\frac{\sin(kr)}{(kr)^3} - \frac{\cos(kr)}{(kr)^2}\right](\mathbf{r} \times \mathbf{M}), \qquad (6.28)
\end{aligned}
$$

where k is a constant and $\mathbf{M}$ can be seen as a constant vector through the central node around which the displacement vectors $\mathbf{u}$ are directed.

6.4   Standing waves between singularities

In the preceding sections I discussed the possibility that there might be a standing wave between a node and a concentric firm shell. *The most important result was that a singularity may form the one endpoint in a standing wave.* An imaginary firm shell is of course not realizable in a spatial continuum of infinite extension. The Universe itself, however, acts like a gigantic cavity from which no waves can escape, so disturbances that were created as a debris from the initial phase of the Big Bang, can never disappear again, but only rearrange or diminish as a result of expansion. On the other hand, singularities are indeed possible entities. The question therefore naturally arises if standing waves can still be formed in the spatial continuum without the participation of any firm shells.

By the identity

$$\sin x \cos y = \frac{1}{2}[\sin(x+y) + \sin(x-y)], \tag{6.29}$$

Equation 6.11 can be rewritten into:

$$\begin{aligned}
\text{div } \mathbf{u} &= \frac{A}{pr}\cos(c_g pt)\sin(pr) \\
&= \frac{A}{2pr}\big[\sin(pr + c_g pt) + \sin(pr - c_g pt)\big], \tag{6.30}
\end{aligned}$$

which can be interpreted as the sum of two spherical waves bouncing back and forth in opposite directions between the firm shell and the center node. The two waves can be said to be reflected successively from the firm shell and the singularity at the center. We realize that standing waves can be treated as the superposition of two or more progressive free waves. This makes the mathematics simpler, and we can investigate more complex systems of standing waves. In this case the waves move with the speed, c_g. The frequency is $f = c_g p/2\pi$, and the wavelength is $\lambda = 2\pi/p$.

Next let us have a look at the energy which is involved in the standing wave between the center node and the shell. At the time $t = 0$ the whole energy is in the div $\mathbf{u}$-field because the $\dot{\mathbf{u}}$-field equals

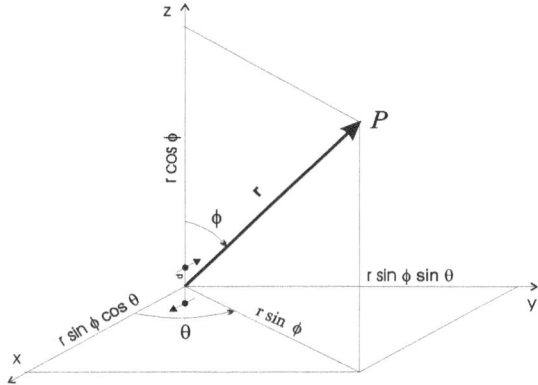

Figure 6.2: Energy around point in space.

zero all over the considered volume. By 6.7 the energy inside a thin shell at the distance r from the center is

$$de = \frac{\lambda + 2\mu}{2}(\text{div } \mathbf{u})^2 \cdot 4\pi r^2 dr,$$

and inside a shell with radius R_n

$$
\begin{aligned}
E_{R_n} &= \frac{2\pi(\lambda + 2\mu)A^2}{p^2} \int_0^{R_n} \sin^2(pr)dr \\
&= \frac{2\pi(\lambda + 2\mu)A^2}{p^2}\left[\frac{R_n}{2} - \frac{\sin(2pR_n)}{4p}\right].
\end{aligned}
\tag{6.31}
$$

We notice that the total energy inside the sphere approaches infinity when R_n grows without any limit. *Hence a single oscillating node cannot exist in the spatial continuum.*

Let us make a new thought experiment with the rigid shell replaced by an ellipsoid, and with an oscillating node in one of the focal points. Then a ray of the spherical waves emitted from that point will travel outwards until it reaches the rigid shell were it is reflected in the direction of the other focal point. The travelled distance before it arrives at that point is independent of the direction of emittance, and it is like the length of the major axis. Now, let the distance between the focal points be like half a wavelength and the major axis like an odd number times the wavelength. Then the wave will reach the other point in opposite phase to when it was emitted. As we have seen the incoming spherical wave is reflected, and since the phase shift is the same at both the rigid sphere and the focal point, the two reflections leave the resulting phase shift unaffected. The wave, when first started, will bounce back and forth between the two focal points forming a standing wave pattern. The oscillations at each point will be in opposite phase to each other.

Around the two nodes we can find the velocity field by superposing the two waves, and in the near vicinity the streamline pattern will resemble what we find around a dipole of a sink and a source. At some distance, however, the pattern becomes more complicated. We can find the total energy of the system by integrating the potential and kinetic energy all over the cavity. If the total energy should happen to be finite when we expand the major axis towards infinity, we could get rid of the rigid ellipsoid, and the energy would just bounce back and forth between the two entities. I'll try to check that out.

First I shall only consider the energy in a volume element dV at a distance r from the center that is great in comparison to the distance $2d$ between the nodes (see Fig. 6.2). Then the angle between the axis through the two nodes and the direction towards the volume element can be considered to be the same, namely ϕ, from both nodes and the center point. The divergence in the volume element at P is the superposition of the divergence from the two nodes at the time $t = 0$

$$
\begin{aligned}
\text{div } \mathbf{u} &= A\left[\frac{\sin p(r + d\cos\phi)}{p(r + d\cos\phi)} - \frac{\sin p(r - d\cos\phi)}{p(r - d\cos\phi)}\right] \\
&\approx \frac{A}{pr}\left[\sin(pr + pd\cos\phi) - \sin(pr - pd\cos\phi)\right] \\
&= \frac{A}{pr}\cos(pr) \cdot \sin(pd\cos\phi).
\end{aligned}
$$

The energy in a zone between $R_m \gg d$ and R_n when $n \gg m$ in the two half spheres is

$$
\begin{aligned}
E &= \int\int\int \frac{\lambda_s + 2\mu_s}{2} (\mathrm{div}\,\mathbf{u})^2 dV \\
&= C\int_{R_m}^{R_n}\int_0^\pi\int_0^\pi \frac{1}{r^2}\left\{\cos^2(pr)\cdot\sin^2(pd\cos\phi)r^2\sin\phi\,d\phi\,d\theta\,dr\right\} \\
&= C\int_{R_m}^{R_n}\cos^2(pr)dr\cdot\int_0^\pi\sin^2(pd\cos\phi)\sin\phi\,d\phi\int_0^\pi d\theta \\
&= C\int_{R_m}^{R_n}\cos^2(pr)dr\left|-\frac{\cos\phi}{2}+\frac{\sin(pd\cos\phi)\cos(pd\cos\phi)}{2pd}\right|_0^\pi|\theta|_0^\pi
\end{aligned}
$$

$$
E = C\left[1 - \frac{\sin(2pd)}{2pd}\right]\int_{R_m}^{R_n}\cos^2(pr)dr, \tag{6.32}
$$

where C is some constant.

The property $\sin(2pd)/2pd$ has a maximum value of unity when $2dp$ approaches zero, while the integral grows beyond any limits when r grows, hence the total energy has no upper bounds except when either d or p approaches zero. *We can conclude that a single isolated dipole cannot exist because it would take an unlimited amount of energy to keep it going.* Notice by the way that E is vanishing in the equatorial direction from the origin and increases when ϕ shrinks towards zero (i.e. it has its maximum value in the z-direction). If we suddenly removed the rigid ellipsoid, the energy would bounce back and forth with decreasing intensity for a while, but eventually it would be totally radiated away. If this is a correct assumption, then some of the energy travels directly between the nodes without taking the detour via the rigid encircling surface.

6.5 Strings of oscillating nodes

The next scenario I want to look into, is a string of nodes oscillating between compression and rarefaction where one node is oscillating in opposite phase to its two adjacent nodes along an unbroken chain of nodes. If we set the distance between the nodes equal to half a wavelength, then the distance between any two oppositely oscillating nodes in the chain will become $(n+1/2)\lambda$ where n is an integer. Like with two nodes at a distance of half a wavelength, the wave sent out from one of the nodes will arrive at the other in opposite phase, and like the example in the previous section, some of the energy will bounce back and forth between these nodes without taking the detour via the rigid surface. It seems like an increasing amount of energy will take the direct route between the nodes as the chain is extended with an increasing number of nodes. At some distance from the chain the superposition of the field from two neighbouring nodes will tend to cancel each other out. The result will be that the field energy will stay almost entirely in the near surroundings of the chain and probably reach its lowest possible level.

If the above assumptions are true, then the myriad of oscillating nodes created in the first fraction of a second after the Big Bang will tend to group into chains of oscillating nodes, and the chains will be very strong and almost unbreakable, because breaking the chains will require an almost unlimited amount of energy to be built up around the chain. I can see no other possible force that could bind together elementary particles - whatever they might be like - and hindering them from flying apart. Thus I shall base the whole model of matter on the conception that matter basically is built up of chains of oscillating nodes that again are bound together by unbroken chains of oscillating nodes even if I find it hard to prove it.

So far I have mostly considered nodes that oscillate between compression and rarefaction mostly because it gives the simplest intuitive picture of the general idea. It should however not be anything in the way that the same goes for the solenoidal types of oscillating nodes. Here the central symmetrical displacements have got to be replaced by cylindrical symmetry around axes in space. For the field from two or more oscillating nodes to be completely eradicated by destructive interference, all the axes have got to have the same orientation in space. That is a necessary condition for the field energy to reach its lowest possible energy state. Thus in any system of oscillating nodes all the spin axes will tend to be pointing in the same spatial direction, so one axial direction stands out as the main direction. The rotations can either be said to be pointing *up* or *down* along this direction.

6.6 Coupled oscillations

In this section I will discuss the possibility that the two basic types of oscillation – spin right/left and compression/rarefaction – can occur in the same set of nodes. In the Navier-Cauchy representation the two wave equations are independent of each other and the fields cannot interact, but not so in the Navier-Stokes representation of a continuous medium. The main reason for this is that the *time derivative* of **u** is not exactly like the *partial time derivative* of **u** with respect on time. In fact [1, p. 116]:

$$\dot{\mathbf{u}} = d\mathbf{u}/dt = \partial\mathbf{u}/\partial t + (\mathrm{grad}\,\mathbf{u})\dot{\mathbf{u}} = \partial\mathbf{u}/\partial t + (\dot{\mathbf{u}} \cdot \nabla)\mathbf{u},$$

and

$$\ddot{\mathbf{u}} = d^2\mathbf{u}/dt^2 = \partial\dot{\mathbf{u}}/\partial t + (\dot{\mathbf{u}} \cdot \nabla)\dot{\mathbf{u}}$$

In the linear theory of elasticity, where we only deal with almost infinitesimal deformations, the difference is insignificant, but in situations where the speed of displacement of nearby points are different, the effect can no longer be neglected. This happens in the near vicinity of the singularities, and here we can find a transfer of energy between the two types of fields in question. Hence the two oscillations may be coupled to each other, and we must expect to find some *normal modes* of oscillations. Finding a mathematical description of those modes can only be obtained by applying the full power of the Navier-Cauchy equation (see Fig. 6.3)[4]. This equation cannot any longer be split up into two separate wave equations, which makes the equation much more complicated. Instead I shall try a more practical approach.

---

[4]Note that the Navier-Stokes equation for a fluid flow has this extra term included.

The complete N-C equation:

$$(\lambda_s + 2\mu_s)\text{grad div }\mathbf{u} - \mu_s\text{curl curl }\mathbf{u}$$

$$= \rho_s\left[\frac{\partial^2\mathbf{u}}{\partial t^2} + (\dot{\mathbf{u}}\cdot\nabla)\dot{\mathbf{u}}\right],$$

which by the identity:

$$\text{grad}\,(\mathbf{A}\cdot\mathbf{A}) = 2[\mathbf{A}\times\text{curl }\mathbf{A} + (\mathbf{A}\cdot\nabla)\mathbf{A}],$$

becomes:

$$(\lambda_s + 2\mu_s)\text{grad div }\mathbf{u} - \mu_s\text{curl curl }\mathbf{u}$$

$$= \rho_s\left[\frac{\partial^2\mathbf{u}}{\partial^2 t} + \frac{1}{2}\text{grad }\dot{\mathbf{u}}^2 - \dot{\mathbf{u}}\times\text{curl }\dot{\mathbf{u}}\right].$$

Figure 6.3: The complete N-C equation.

Consider a simple harmonic oscillation that might occur. Let it be a coupled oscillation between rotation around axes through each of the nodes in a chain of oscillating nodes and movement to and from the nodes. The axes are supposed to be normal to the chain and in the same plane. The rotation can be said to alternate between the right/left direction and compression/rarefaction during a period. Say that left rotation is coinciding with compression, then right rotation is coinciding with rarefaction, and vice versa.

Now let us consider a tiny concentric ring with mass m_i and radius r_i around one of the axis when the rotation ω_i is at its maximum starting to build up a peak right torque. The *angular momentum* of the ring is given by

$$\begin{aligned} L_i &= I_i\omega_i = r_i m_i v_{[\text{tang}]}, \qquad v_{[\text{tang}]} = \omega_i r_i, \\ &= m_i r_i{}^2 \omega_i. \end{aligned}$$

The *law of conservation of angular momentum* states that when no external torque acts on an object or a closed system of objects, no change of angular momentum can occur. If a torque τ_i is present, we have the relation:

$$\tau_i = \frac{\partial L_i}{\partial t}.$$

At the moment when the rotation of the ring is at its maximum, however, the torque is zero and we can set up the relation:

$$\frac{\partial L_i}{\partial t} = 0,$$

$$m_i \left(2r_i \dot{r}_i \omega_i + r_i^2 \frac{\partial \omega_i}{\partial t} \right) = 0,$$

$$\frac{\partial \omega_i}{\partial t} + 2\frac{\dot{r}_i}{r_i}\omega_i = 0,$$

$$\omega_i = \frac{1}{r_i^2} + C.$$

When the radius of the ring is decreasing, the rotation is increasing.

The rotational energy in the ring is given by:

$$E_i = \frac{1}{2}I_i\omega_i^2.$$

Clearly energy is transferred from irrotational to rotational energy during compression or rarefaction and back again when the tension is released.

If the amplitude of rotation is equally strong in both directions, then the mean pressure over some periods will be like the normal background pressure in the spatial continuum, but consider an oscillation where the rotational amplitude in just one direction is increasing. Then the mean pressure is affected. Say that the amplitude to the left is increasing, and that rotation to the left is coinciding with compression. Then the compression in the node in question is increasing in step with rotation. On the other hand, if rotation to the left is coinciding with rarefaction, then rarefaction is enhanced. This asymmetry is supposed to play a role in explaining what electrically charged particles are like.

Chapter 7

Matter as strings of oscillating nodes

One of the most intriguing questions in science is how an entity can be both point-like and wave-like at the same time. It is questions like this I will address in this chapter. In Chapter 6, *'The oscillating node'*, I discussed the possibility that standing waves could occur between singularities in the spatial continuum, and found that such singularities would tend to group into long chains of nodes, each node in the chain oscillating in opposite phase to its two neighbouring nodes. The spatial mass displaced from one node at a given time would be found in the neighbouring nodes such that the net displacement from such a string would be zero. In a chain of nodes with rotational deformations, the spin in the entire chain would sum up to zero. In this chapter I will extend the discussion to encompass also chains with a certain displacement and spin.

7.1 High and low pressure nodes

As we have seen, space is taken to be an elastic continuum of infinite extension in which there may be singularities in the form of oscillating nodes, which oscillate between compression and rarefaction along an infinitely long chain of nodes. Compression and rarefaction are located to the area around each node, but for convenience I shall speak of the spatial mass as if it was going into or out of the nodes themselves like sinks and sources. Normally the spatial mass that is displaced from one node goes as a compression into its adjacent nodes such that the net displacement is zero, but there may be a single node in the chain that is exited to a higher degree than what is taken up by the other nodes in the chain. It may be a pressurized node which can take up an amount of spatial mass, or it can be an evacuated node which can be thought of as a bubble.

Let us consider an evacuated node, which I occasionally will refer to as a *bubble*. The bubble is of course not stable, but will immediately be filled up with an inward stream of spatial mass from say the next node in a chain of nodes. When that node in turn is evacuated to the same level as the former node, it will collapse sucking up spatial mass from the next node, and so on. In that way the bubble will be moving in a step-wise manner from node to node along the chain. When the domain around the node representing the bubble is exactly filled up, the inward stream of spatial mass has its maximum speed and will continue for a while until compression stops the movement and reverses the stream. Therefore nodes that are left behind have got to oscillate with decreasing amplitudes as the bubble recedes. On the other hand the next node in the chain cannot suddenly be inflated to the exited level, but has got to start oscillating with increasing amplitude in good time before the bubble arrives. Hence the arrival of the exited node has got to be preceded by a pilot chain of nodes that oscillate with increasing strength as the inflated node approaches, and followed by a trailing chain of nodes with decreasing strength as the bubble moves forward. The entire chain of nodes may extend far out in both directions.

This picture conforms well with the findings in Chapter 6 where I found that oscillating nodes only are possible in an infinitely long chain of oscillating nodes. Here the assumption has got to be modified to incorporate very long, steadily increasing strings of nodes (which I will dub the *preamble*) that culminate in a strongly inflated node (dubbed the *bubble*) whereupon the trailing nodes (which I will dub the *postamble*) gradually decreases towards zero. In this picture all matter has got to be viewed as composed of long, almost infinitely long strings of oscillating nodes with a strongly excited node at its core.

A function of the type $\sin(\omega t)/t$ seems to cover this situation well for the oscillation of each node along the chain of nodes when t varies between $-\infty$ and $+\infty$[1]. This is a calculated guess and I cannot prove it, but I will try to see what implications it will have. The limit of the function as t approaches zero is given by

$$\lim_{t \to 0} \frac{\sin(\omega t)}{t} = \omega,$$

where ω is the angular velocity of the oscillation. (See the leftmost graph in Fig. 7.1.) Let us give each node along the chain (represented by a black dot in the graph) an integer number n such that the next node in the chain has the number $n + 1$, and let the displacement from the domain around node #n at the time t be given by:

$$D_n(n, t) = k \frac{\sin[\omega(t - n\pi/\omega)]}{t - n\pi/\omega}, \qquad (7.1)$$

where k is some constant, $n = \{ \cdots -1, -2, 0, 1, 2, \cdots \}$. The displacements from each node is represented by the vertical columns in Fig. 7.1.

First we notice that according to the above assumption, when rotation is disregarded, the displacement from node #0 at the time $t = 0$ is given by:

$$D_0(0, 0) = k \lim_{t \to 0} \frac{\sin(\omega t)}{t} = k\omega$$

while the displacements from all the other nodes in the chain are zero, i.e. the displacement from the entire chain of nodes is $k\omega$.

---

[1]The sinc(t)=sin(t)/t function resembles a Dirac's delta function, which is one at $t = 0$ and zero everywhere else, when one levels out the oscillatory components. Hence a function of the type, $f(x) = D \cdot \sin(x - vt)/(x - vt)$, can be thought of as a bubble with a displacement, D, moving with velocity, v, along the x-axis.

Half a period later node #1 is fully inflated while the displacements from all the other nodes again are zero. In fact this is the case for all times $t = n\pi/\omega$:

$$D_n(n, n\pi/\omega) = \lim_{t \to n\pi/\omega} k\frac{\sin[\omega(t - n\pi/\omega)]}{t - n\pi/\omega} = k\omega. \qquad (7.2)$$

In the intermediate phases, the displacement from the exited node gradually decreases while the new displacement builds up in the next node. In this way the inflated node moves from node to node with a frequency like $2f$ while all the other nodes along the chain oscillate between fully inflated state and fully compressed state with the same frequency. The blue columns in Fig. 7.1 show the displacement from each node in successive steps in the course of half a period. They are in fact the graph of a function given by 7.1.

If this shall be a viable model, however, then at least the net displacement has got to be the same during the intermediate phases, so we write down the sum of the displacements from all the nodes in the chain at any time

$$\begin{aligned} D(t) &= \lim_{n \to \infty} \sum_{m=-n}^{n} D_m(t) \\ &= \lim_{n \to \infty} \sum_{m=-n}^{n} k\omega\frac{\sin(\omega t - m\pi)}{\omega t - m\pi}, \qquad \omega t - \pi m \neq 0. \end{aligned}$$

First we notice that $\sin(\omega t - m\pi) = \sin(\omega t) \cdot (-1)^m$ for any $m \in \{\cdots -2, -1, 0, 1, 2, \cdots\}$. Hence:

$$D(t) = k\omega \sin(\omega t) \lim_{n \to \infty} \sum_{m=-n}^{n} \frac{(-1)^m}{\omega t - m\pi}. \qquad (7.3)$$

From [20] I fetch the limit representation of $\csc(z)$:

$$\csc(z) = \lim_{n \to \infty} \sum_{m=-n}^{n} \frac{(-1)^m}{z - \pi m} \qquad /; \frac{z}{\pi} \notin Z,$$

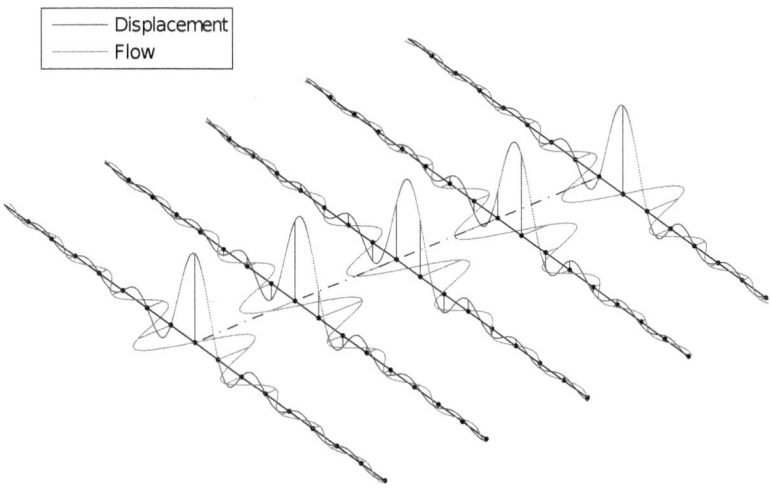

Figure 7.1: Model of bubble.
Model of a "bubble" and its associated nodes in 5 successive steps separated by one eighth of a period. The vertical columns represent the displacement and the horizontal columns the flow to and from each node in the string.

where $\csc z = 1/\sin z$. Hence

$$D(t) = k\omega \sin(\omega t)\frac{1}{\sin(\omega t)} = k\omega. \qquad (7.4)$$

We can conclude that the displacement from the whole chain of oscillating nodes is the same at any time during an oscillating cycle. A chain of nodes meeting these requirements will represent a net displacement, which we could call a bubble, moving with some speed v through space.

We also notice that the displacement is given by a constant times the frequency. If there is a one to one correlation between displacement and energy, then the energy is given by some constant times the frequency, just like the Plank-Einstein relation, $E = hf$. In Sec. 5.1 I discussed a naïve model of matter where I found that there should be a fixed relation between energy and displaced spatial mass given by the relation

$$D = \frac{E}{3(\lambda + 2\mu)}.$$

(See Fig. 5.1.) If we assume that this is a general principle also the other way around such that to a given displacement there is a net amount of energy given by

$$E = 3D(\lambda + 2\mu) = 3k(\lambda + 2\mu)\omega = \kappa\omega, \tag{7.5}$$

where κ is a temporary constants given by $\kappa = 3k(\lambda + 2\mu)$, then

$$k = \frac{\kappa}{3(\lambda + 2\mu)}. \tag{7.6}$$

Like Plank's constant, $\hbar$, the temporary constant κ relates to the energy of a material particle, hence there should be a fixed relation between the two constants.

We now turn back to the amplitudes of the oscillation by each node, which by the way are situated at a distance of half a period from each other. Between maximum compression and inflation there has got to be a phase of inward and outward flow of spatial mass to and from the nodes. The rate of change of the displacement from each node will naturally be a measure of the flow, so by taking the time derivative of the displacement we will get an expression for the flow, i.e. the flow through a surface around the node where $\operatorname{div} \mathbf{u} = 0$. It turns out that the flow may be represented by a function of the type $\frac{\omega}{t}[\cos(\omega t) - \frac{\sin(\omega t)}{\omega t}]$, or for the entire chain of nodes the flow, $F(n, t)$, from node $\#n$ at the time t is given by

$$F_n(n, t) = \frac{h\omega}{t - \frac{n\pi}{\omega}}\left[\cos(\omega t - n\pi) - \frac{\sin(\omega t - n\pi)}{\omega t - n\pi}\right]\delta(\omega t - n\pi).$$

We can draw a graph of the displacement from each node along the y-axis and the flow out of each node along the z-axis. Just think of the nodes as not moving entities and the graph as moving forward along the x-axis just showing the displacement and flow from each node as a function of time. The first graph in Fig. 7.1, page 139, shows a moment when the displacement is from just one of the nodes, while the next tree graphs show the situation 1/8, 2/8, and 3/8 of a period later, and the last graph shows the displacement when the next node is fully inflated 4/8 of a period later. We notice that the excitation reaches its maximum value two times in the course of one cycle of oscillation.

For symmetrical reasons exactly the same development can be done for a compressed node moving along a similar string of nodes, but then the displacement is the negative of D, i.e. a compression.

This idealized model can possibly explain some of the properties of material particles if the path is curled up in some way. It can account for both the wavelike and the particle-like property of matter, but not for the property of spin. Spin, however, characterizes the great group of elementary particles called Fermions with half integer spin, that are bound together by Bosons with integer spin, to form all kinds of matter. Hence possible particles as described above can, if they exist, at best be related to dark matter.

7.2   The photon

Photons are known to possess spin and electromagnetic properties like polarization, which the hypothetical spin-less particles in the previous section do not have. In order to meet these requirements, it will be necessary to consider strings of nodes that oscillate between torsion in opposite directions in stead of nodes that oscillate between compression and rarefaction.

Consider a straight string of nodes oscillating between rotation in the right and left direction such that one node oscillates in opposite phase to its two adjacent nodes. The field around the nodes can be fully described by the displacement field, $\mathbf{u}$, and the velocity field, $\dot{\mathbf{u}}$. The field components generated by each node form concentric circles around the rotation axes through the node, and the resulting field is the superposition of the components from all the nodes. Since the axes is supposed to point in the right/left direction, they will all be parallel, and all the displacement components and the resulting field components will be in planes normal to these axes. By defining a Cartesian coordinate system with the z-axis pointing in the direction of the rotation, the field may be fully described by the two components in the xy-plane. Alternatively one can entirely suppress the z-axis and represent the field by the real and imaginary parts of a complex number. This will make the treatment of the displacement field much simpler (see Sec. 7.8).

Such a string of oscillating nodes resembles a highly directional Yagi-Uda antenna that is known to collect radio waves into a narrow beam. As with a Yagi-Uda antenna, the field energy at some distance to the side of the string is suppressed by destructive interference while being reinforced by constructive interference closer to the string. In the same way as the oscillation of the elements in the antenna generate electro magnetic waves, the preamble of the string should generate a progressive solenoidal wave that is supposed to follow the movement of the exited node. If we study the whole chain of nodes, however, we see that the nodes in the postamble oscillate half a period out of phase with those in the preamble (see upper graph in Fig. 5.4 on page 88) and would tend to kill the wave generated there.

If there on the other hand should be a flip-over between the left/right direction of the built up torsion in the exited node, then the two waves would reinforce each other (see lower graph in the same figure). A flip-over of torsion from one direction to the other can perhaps occur if there is some kind of instability when the node reaches its exited state. Say that the peak torsion is released into a rapid spin that again builds up a torsion in the opposite direction. From there on, the torsion is oscillating back to nil in the postamble. Since this happens over again each time a new node reaches its exited state, the whole system will perform an ongoing spin.

Even if the flip-over might happen in other ways, for example if the torsion is wrenched over from the right to left direction by passing through other directions, the spin would necessarily be the same. Such a flip-over resembles an interesting phenomenon that frequently is occurring in physics, namely that of *spinors*. Discovered by Èlle Cartan in 1913, it has among other things played a mayor role by studying the intrinsic spin properties of fermions. One popular example of a spinor is the Balinese cup trick[2]. A Balinese dancer takes an open cup partly filled with water or wine and turns it 720^0 around, or a multiple thereof, with her feet placed firmly on the floor, without spilling a drop of water or loosening the grip of the cup. It is a remarkable phenomenon because one should think it impossible to turn something repeatedly around when holding the object in a firm grip and keeping ones feet planted on the ground. Demonstrations of the trick can easily be found on the internet and can be performed by anyone. It goes on in two stages: First the cup is turned one full revolution of 360^0 with the arm in a lower position, and then the rotation is completed with another full turn in an upper position. During this process the cup is rotating with a given *spin angular momentum*, or just *spin*.

The spin of a rotating body is given by $L = I\omega$, so the energy can also be expressed as $E = L\omega$. We obtain the relation between energy and spin of the system given by:

$$L = \frac{E}{\omega}.$$

If the energy of the photon is given by $E_p = \hbar\omega$, then the spin of the photon is:

$$L = \hbar,$$

as expected.

If this approximates the correct picture of a photon, then it should appear as shown in Fig. 5.5 on page 90. The black graph shows the displacement around each node (black dots), and the

---

[2]It comes in a lot of variants: The plate trick (also known as Feynman's plate trick), Dirac's belt trick, Dirac's string trick, spinor spanner, Bredon high-five, or quaternionic handshake.

coloured/gray graphs show the generated progressive wave. The graph shows the torque around each node at the exact moment when one of the nodes reaches its exited state. We see that the toque around all the other nodes then are exactly like nil. Hence all the potential energy at that moment is in the torsion around the exited node, while the kinetic energy is divided between all the other nodes. We can assume that in a freely oscillating system the energy is equally divided between the two field, so the total energy is the double of the potential energy. When the torque is released into rotation, it will represent half of the total energy. Thus the spin is directly related to the energy of the system.

This completes the somewhat naïve picture of a photon. A rotational deformation around an axis through a singularity moves in a stepwise manner from node to node along a string of oscillating nodes with the propagating speed of solenoidal waves. Its particle-like nature is attended to by the highly exited node at the core of the system, and the chain of oscillating nodes creates around itself a progressive transversal wave that follows the system and accounts for its wave-like nature and polarization. The condition that the oscillations in the preamble have got to be in phase with those in the postamble, forces the torsion in the exited node to flip between opposite directions initiating a process related to the behaviour of spinors. It gives the system a spin, probably along the axis of movement, that has a fixed value independent of the oscillatory frequency of the system. And finally, the energy is fixed in relation to the frequency of the system making the energy in a beam of photon to appear only in a distinct amount of quanta, namely photons.

A photon is also known to appear in a state of elliptical or circular polarization. In the approach above I suggested that the rotation axes are normal to the string of nodes. This is probably the normal state of a photon, but the condition that all deformations can be represented by only two components will be met also when the axes are pointing in other directions, but remain parallel. This opens for the possibility that there may be components of **u** along circles around the string, even to the extreme that they are all directed along the chain itself. Thus circular polarization might be a state where the rotational axes are directed along the chain itself.

7.3 Virtual photons

In the section above I tried to imagine what a real photon is like. Photons, however, are also known to appear in a virtual state that plays an important role in describing electric fields.

Imagine a photon with a preamble like the one described above that does not flip over the torque from one direction to the other. Following a single node in the chain, we would find that in the preamble it is building up a torque that is released in the postamble. Such a chain of nodes would carry with it a torque that is not released into a spin, but all the same it has a *potential spin* that may be released into a real spin if it interacts with other systems.

Like the real photon it will try to generate a progressive wave in the preamble, but the wave is half a period out of phase with that tending to be generated in the postamble. This will appear like an internal conflict that makes the whole chain unstable, so, if a string like that should be created in some way, it would immediately tend to break down again. Hence it would be very short-lived and undetectable, so we dub it a *virtual photon*.

Say that there in addition is a coupled oscillation, as discussed in Sec. 5.5 on page 90, between rotation and compression/rarefaction in the preamble of the chain. While rotation alternates between the left and right direction, pressure alternates between compression and rarefaction. Let rarefaction be coinciding with left rotation and let the virtual photon build up to a left torque. Then an underpressure is built up in the preamble leading to an evacuated state in the exited node, which we could call a "bubble". Hence the virtual photon would carry with it a bubble. A virtual photon building up a right torque, would accordingly bring with it an extra amount of spatial mass. Hence a virtual photons can carry with it either a hole or an extra lump of spatial mass, and we could say that it is either down or up polarized.

It is probably impossible for a single bubble or a single extra lump of matter to suddenly come into existence. But it seems reasonable that two such entities can be created by pair production and disappear by annihilation. The spatial mass needed to create the extra lump could be taken from the bubble. I will look at the energy accounts to make it possible in the next section.

7.4   Electric fields

In this model of space and matter, an electric field can be represented by nothing but a flow of spatial mass. I have found it more likely that the electric field is positive in the direction of the down particle flow, but that turns out to depend more on convention than necessity. A steady flow of spatial mass in an elastic continuum is hard to imagine, but the proposed model of space and matter stands and falls with the assumption that such a field can be realized.

Say that we have an area of space that is not empty, but is filled with up and down polarized virtual photons that is created by pair production and disappear by annihilation all the time. For their short existence they are all thought to move in opposite directions, say that the down particles move to the left and the up particles to the right.

Imagine a bubble in a liquid moving a short distance from A to B during the time Δt. That is equivalent to saying that the hole disappears from around A and reappears around B. To accomplish that, the liquid that originally was found around B has got to be transported back filling up the hole at A. The transport has got to be as a dipole flow starting i B and ending in A. On the other hand small compressions – or lumps – of liquid moving in the opposite direction, say from C to D, would create a dipole flow from C to D, i.e. in the same direction as the other dipole.

Since space presumably is filled up with virtual photons moving just like that, there will be lots of dipoles within a certain area, and by adding up all the flow components, they will produce a positive velocity field to the right corresponding to an electric field directed from right to left in the area. Electric fields have an intrinsic amount of energy depending on its strength. In this representation the energy is divided between the energy that goes into the virtual photons and the flow fields between them. Finally, what is needed to keep such a field going are two engines, one that spews out up polarized and suck in down polarized virtual photons, and a another one that works the opposite way around.

7.5 The electron

Leptons are a group of elementary particles with half integer spin. The first generation is the electronic leptons, comprising the electron, e^-, and electron neutrino, ν_e; the second is the muonic leptons, comprising the muon, μ^-, and muon neutrino, ν_μ; and the third is the tauonic leptons, comprising the tau, τ^-, and the tau neutrino, ν_τ. All these particles come with their antiparticles, the most famous of which is the antiparticle to the electron, namely the positron. Models of electrons and positrons must be expected to be much more complicated and difficult to explain than that of a photon, but I'll give it a try.

The most prominent feature of electrons and positrons are their electrical charge. In this model of space and matter, electric charges can only be explained as sources and sinks in the elastic continuum. First, it looks like an impossibility that spatial mass can be created in one point and disappear in another almost as if there should be a loophole between the two entities, but it turns out that it is precisely what we have got to be looking for.

It is known that under certain conditions a photon with sufficiently high energy can split into an electron and a positron. The two new entities can even go into a short lived composite particle with the electron and positron bound together in an exotic atom known as a Positronium (Ps) [17] (see also Fig. 7.2). The photon has integer spin, i.e. it is a Boson, and since spin is a conserved property, the two new entities should end up with only half integer spin each, i.e. they are Fermions. The photon moves with the speed of light while the two new leptons can move with any speed

Positronium

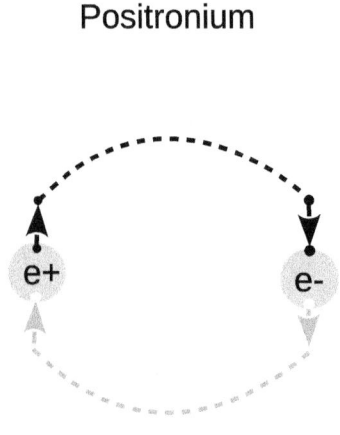

Figure 7.2: Positronium.
A positronium is a short lived particle consisting of an electron and a positron. The illustration indicates that an amount of spatial mass is emitted from the positron and received by the electron, while a 'bubble' evacuated by the same amount of spatial mass is transported the other way. The dashed lines indicate that the transport is performed by free virtual particles.

lower than that. We have got to assume that on a somewhat unpredictable manner the two chains of nodes now takes the form of curved paths, which the two exited nodes have got to follow.

When a photon – as seen in this model – splits up into two new entities, not only spin has got to be conserved. If one object displace spatial mass, that mass has got to be squeezed into the other object. The total mass must be conserved. Since a photon is supposed to displace no spatial mass at all, that has got to apply to the two new entities as well. Hence if one of the items is to become a source, then the other has got to become an equally strong sink. This will ensure that charge is a conserved property also in this model.

Even if the strings of nodes in the post- and preamble are being curled up in some ways, we can let the ordinate represent deformation and the abscissa time, and then draw the graph as a straight line in a plane. This will help us to visualize what is going on, so let us follow a single node in the preamble of these two new particles. Say that there is a coupled oscillation between left/right torque and compression/rarefaction in the preamble of the leptons respectively. In the course of one cycle, one of the nodes will build up an increased pressure while the other will gradually be evacuated until they both reach the point of excitation. The first will act like a sink and the other like a source while they are in the preamble. At the point of maximum excitation, torque is supposed to flip over to the opposite direction while the bubble and the compressed lump of spatial mass are torn loose and travel on as separate entities. After the flip-over the nodes are supposed to capture a virtual particle of opposite polarization to the one they emitted. The node that built up and emitted a compressed lump of spatial matter will capture a bubble that gradually is filled in the postamble continuing to act like a sink, and vice versa.

The node that built up and emitted a compressed lump of spatial matter in the preamble will capture a bubble that gradually is filled up in the postamble continuing to act like a sink in its entire time of existence, and vice versa. The same will happen to all the nodes in their respective particles. One will act like a sink and the other like a source just as we expect two charged particles to behave. The virtual particles will act like the loopholes that bring spacial matter back from the sink to the source by the mechanism describing the electric field.

If the virtual particles carry with them an element of torque as proposed for virtual photons, or they just are compressed and evacuated objects as discussed in Sec. 7.3 is hard to say. For the time being I prefer to use Occam's razor an suppose that the virtual particles consist of only compression or rarefaction, but nevertheless leave it as an open question[3].

7.6 Quarks

Hadrons are particles that are composed of a quark-antiquark pair (mesons), or three quarks (baryons). The only stable hadrons are the proton and the neutron. The latter is only stable within the nucleus. Electrical forces are far to weak to keep the quarks together in hadrons, so quarks are kept from flying apart by the *strong forces*,

---

[3]Occam's razor as he formulated it: "Numquam ponenda est pluralitas sine necessitate" [Plurality must never be posited without necessity]

which are about 137 times stronger than electromagnetic forces at distances of about 10^{-15} meter – the size of a nucleus.

Think of quarks as engines like the leptons, that continually spew out virtual particles of one polarization and swallow virtual particles of the opposite polarization. A quark and its anti-quark are two particles that are a mirror image of each other. If one emits up polarized and receives down polarized virtual particles, the other will do just the opposite. Consider a meson with two such particles, i.e. a quark/anti-quark pair, close to each other. They would tend to annihilate, but let them for a short time exist as separate entities. Then the up polarized virtual particle emitted from one of the two entities could immediately be caught up by the other particle, and the down polarized counterpart from the second particle could be caught up by the first. Unlike the binding force between leptons, where free virtual particles make up an electric field, the two new particles make a direct connection like two ropes that tie them together. This is not a one time event. The connection is renewed every time a node in the chain reaches its exited state. The strength of the binding is entirely dependent on the strength of the chain of nodes between the two entities. In Sec. 6.5 I discussed the strength of such chains, and found that they probably are extremely strong. Therefore leptons and quark divide from each other in that the first are bond together by an electric field maintained by means of *free virtual photons* while the others are bond together by gluons that are *unbroken strings of nodes* attached to a quark at both ends.

While mesons are unstable particles that can only exist for a short period, baryons may be much more stable. A free neutron has a half-life of about fifteen minutes while discussion goes if a proton is everlasting. Baryons are made up of three real quarks bound together by gluons, but in addition there are virtual quarks that complicates the picture. Here I'll try a somewhat naïve approach to see if some of the properties in the Standard model have any counterparts in the present model.

Consider the most stable baryon, namely the proton. In the Standard model it is considered to be made up of three quarks: Two up-quarks with electric charge $+2/3$, and one down-quark with

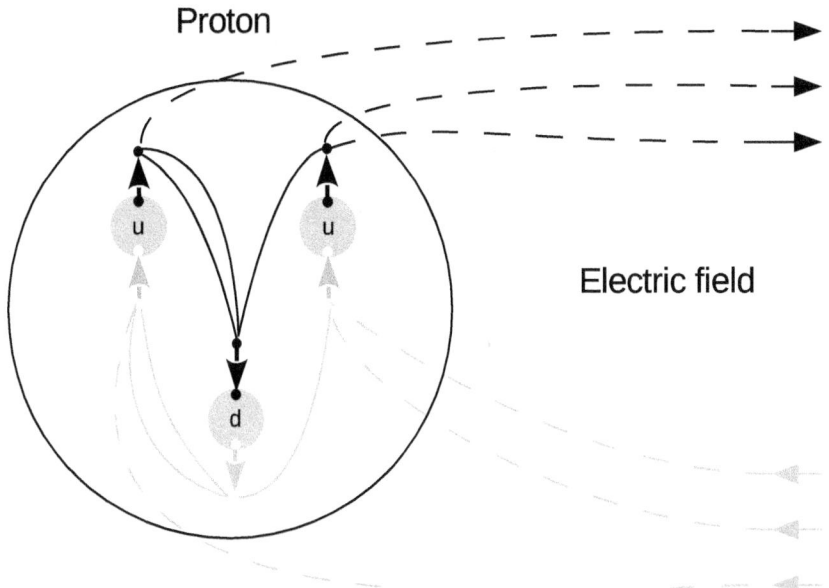

Figure 7.3: The proton.
The up-quarks send away an amount of spatial mass and receive a bubble that is short of exactly the same amount of mass, indicated by black and gray arrows respectively. Stippled lines show the paths of virtual photons and solid lines of gluons. A black line indicates that the particle is up and a gray line that it is down polarized. Each line show the transfer of ±spatial mass corresponding to one third of the charge of an electron.

electric charge -1/3. Together thy add up to one unit charge. The quarks also have color charges that add up to white, and they are bound together by the strong forces mediated by gluons (see Fig7.6). The three quarks are not allowed to to reach their excited stage at the same time, but have got to send out and receive the exchanged volume of spatial mass in succession indicated by black and gray arrows in the Figure. Now, consider the two o'clock quark. It sends out 2/3 of its emitted mass as free virtual photons, and 1/3 as a gluon destined to reach the down quark positioned at six o'clock when that quark has its turn to make the transition. On the other hand the two o'clock quark receives a full 'bubble'; 2/3 of it in the form of free virtual photons from the electric field and 1/3 of it as a negatively polarized gluon that were sent out from the six o'clock quark at an earlier time. In the same way the two next quarks follow suit in succession until a full cycle is reached. Then the proton has sent out up and down polarized free virtual photons corresponding to the unit charge of an electron and exchanged the same amount between them along unbroken chains of nodes between the quarks.

The QCD model ascribes color to quarks and gluons. If we distribute all colors of the rainbow around the clock, then each quark can be ascribed a color depending on when it reaches its point of excitation. If movement goes in fits and starts, the time of excitation may wobble back and forth between the two other colors and thus make a color change after each event. There is a time difference between when a ±mass starts it travel along the chain of nodes in a gluon and when it reaches its destination. Hence a gluon has got to be ascribed two different colors. Hadrons are also known to have different energy states called resonances. Gluons have tips and a tails that are at cut off points at different phases (colors) where they connect to two quarks with different colors, but the colors of the cut off points do not change if we put in a whole number of wavelengths in between. So even if a chain of nodes is not stretchable, it may still have different lengths corresponding to the number of rungs it consists of. Such differences may correspond to different energy states called resonances.

Even if there according to this proposition, also has got to be some electric attraction between quarks, the strength of this force is minute in relation to the binding strength of the unbroken chain of nodes. Mass is transported from the plus charge to the minus charge along a mechanism that is similar to a bucket brigade, the pair productions are the places where full buckets are exchanged with empty ones. Gluons are more like double pipelines where water is transported one way and the displaced air the other way. In reality, however, both do the same job; they transport something (spatial mass) from one place to another. Finally, links between

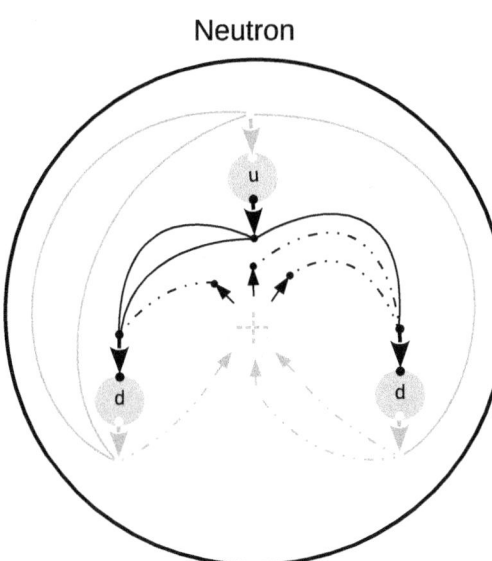

Neutron

Figure 7.4: The neutron. *The neutron has two down quarks and one up quark and no electric charge as seen from the outside. Still a full unit (here represented by three lines) of spatial mass has got to be exchanged between the quarks through the gluons and virtual photons. To obtain that, it has got to posses an intern charge distribution that implies a positive center charge. Notice that the arrows represent exchange of spatial mass and not electric field lines.*

quarks need not only be between quarks in the same hadron. The gluons may occasionally link up to a quark in another hadron, and it is those stray connections that have got to account for the forces tying baryons together into a nucleus.

It is a fascinating thought that all boils down to one single elementary particle and its anti particle. All the other particles come as minor variations of this primordial particle. Well, this is perhaps drawing it too far, but let us instead try to figure out how a neutron fits into the picture. The neutron consists of three quarks, two down quarks and one up quark. It has no electric charge so all exchanges of spatial mass must be kept within the neutron. Since the amount of spatial mass emitted and received by each of the three quarks presumably shall be the same, it is impossible to see that it can be achieved by a number of direct exchanges. However, if there within the boundary of the neutron is a virtual positron, the calculation may work out to be correct (see Fig. 7.4). This is in accordance with Quantum chromodynamics (QCD), which indicates that that is just what we can expect[4].

The best picture of a quark I can think of, is to resemble a quark with an octopus with tentacles pointing out in all directions seeking after other tentacles to make a handshake with. Thus the body of the octopus is the *quark*, and the tentacles are the *gluons* that bind quarks together into a hadron. Residual tentacles may stretch out from the hadrons and bind them together into an atomic nucleus.

Mass is transported from the plus charge to the minus charge along a mechanism that is similar to a bucket brigade, the pair productions are the places where full buckets are exchanged with empty ones. Gluons are more like double pipelines where water is transported one way and the displaced air the other way. In reality, however, both do the same job; they transport something (spatial mass) from one place to another. And moreover, the basic transported quanta are exactly the same for leptons and quarks.

---

[4]NEUTRON CHARGE DISTRIBUTION AND CHARGE DENSITY DISTRIBUTIONS IN LEAD ISOTOPES S. Haddad and S. Suleiman, Physics Department, Atomic Energy Commission of Syria, P.O.BOX 6091, Damascus, Syria.

"Quantum chromodynamics (QCD) calculations indicate that the neutron has a positive core and a negative outer region [2]." (They cite 'Jo van den Brand, P. de Witt Huberts, Physics World, February 35, 1996', to which I do not have access.)

7.7 The material body

A material body is a system of particles bound together mainly by strong and electromagnetic forces. The particles are characterized by their excitations with frequencies that are dependent on energy, and the events are not allowed to be congruent. Therefore, the total frequency of excitations increase with the number of particles and thus the energy of the system of particles. It might be of interest to see how a body with these characteristics performs if it moves with a certain speed through space. To get an impression of that, we can first consider the system in a frame that follows the system, and then as it is when seen from a frame that moves with a certain speed in relation to that. I found in Chap. 4 that all basic formulas are invariant by Lorentz transformations, so we can apply these transformations in order to see how the performances changes when going from one coordinate system to another.

Consider a stationary oscillatory system in one frame, (x_0, y_0, z_0):

$$\psi_0 = A(x_0, y_0, z_0)e^{-i\omega t_0},$$

and transform it to another system that is moving with a speed, $v \ll c$, along the x-axis:

$$\psi = A\left(\frac{x - vt}{\sqrt{1 - v^2/c^2}}, y, z\right)e^{-\omega \frac{t - vx/c^2}{\sqrt{1 - v^2/c^2}}}$$

$$\approx A(x - vt, y, z)e^{i\omega \cdot v/c^2(x - c^2/v \cdot t)}.$$

In the moving frame the system represents two wave movements: One with group velocity, v_g, and one moving through the wave packet with a velocity, v_m:

$$v_g = v,$$
$$v_m = c^2/v.$$

The last equation represents the matter wave proposed by Louis de Broglie in 1924. That material particles exhibit wavelike properties have been shown, not only for electrons, but also for bodies the size of a molecule. It is a central property in quantum mechanics as described by the equations of Schrödinger and Dirac.

7.8   The Schrödinger wave equation

In this model of space and matter, material particles consist of strings of oscillating nodes where each node oscillates between rotation in two opposite directions, and where the rotational directions of all the nodes are parallel to each other. The strings are a composite of different entities like leptons, virtual photons, quarks, gluons, etcetera, which are bound together to the group of particles that form a material body. The highly exited node in each chain moves in a stepwise manner along the curled up chain of nodes with the speed of c. The whole group, however, is advancing with a much lower group velocity, $v \ll c$. The field around each node is composed of concentric outwards and inwards progressive waves that interfere with each other forming standing waves, and the complete pattern is made up of a superposition of the waves from each node in the wave packet. Since all the rotational axes are pointing in the same spatial direction, all deformations will be in planes normal to these axes and can be fully described by the two components in this plane. That makes it possible to represent the solenoidal displacement fields as a superposition of the real and imaginary parts of the field vector, $\mathbf{u}$, around each node in the vicinity, which adds up to the complex number $\psi(\mathbf{r}, t)$. The property, ψ, itself can never be observed, but the squared value of ψ, i.e. $\|\psi\|^2 = \psi^*\psi$, is a real property, which is like the square of the displacement vector, $\mathbf{u}(\mathbf{r}, t)$. The unobservable property ψ, then, represents a *state vector*, and in a notation invented by Dirac it can be written $|\psi(\mathbf{r}, t)\rangle$. In this notation the squared value of $|\psi\rangle$ becomes $\||\psi\rangle\|^2 = \langle\psi|\psi\rangle$, where $\langle\psi|$ is the complex conjugated of $|\psi\rangle$.

The displacements outside the singularities can be described by the Navier-Cauchy equation (3.7, p. 40) without any external forces. It is given by:

$$(\lambda_s + 2\mu_s)\text{grad}\,\text{div}\,\mathbf{u} - \mu_s\text{curl}\,\text{curl}\,\mathbf{u} = \rho_s\frac{\partial^2\mathbf{u}}{\partial t^2},$$

which by the identity, $\text{curl}\,\text{curl}\,\mathbf{A} = \text{grad}\,\text{div}\,\mathbf{A} - \nabla^2\mathbf{A}$, can be transformed into:

$$(\lambda_s + \mu_s)\text{grad div } \mathbf{u} + \mu_s \nabla^2 \mathbf{u} = \rho_s \frac{\partial^2 \mathbf{u}}{\partial t^2}.$$

The solenoidal field is described by setting div $\mathbf{u} = 0$, and we obtain:

$$\nabla^2 \mathbf{u} = \frac{\rho_s}{\mu_s} \frac{\partial^2 \mathbf{u}}{\partial t^2}.$$

The divergence-free displacement field around an oscillation node can be represented by a complex number, $\psi(\mathbf{r}, t)$, as stated above, and by inserting the wave speed, $c^2 = \mu_s/\rho_s$, we obtain the wave equation:

$$\nabla^2 \psi(\mathbf{r}, t) = \frac{1}{c^2} \frac{\partial^2 \psi(\mathbf{r}, t)}{\partial t^2}, \tag{7.7}$$

which can be solved by the product method. Notice that I only seek solutions that is a superposition of standing waves around a plethora of oscillating singularities with rotating axes all pointing in the same direction as described above.

Let $\psi(\mathbf{r}, t) = g(\mathbf{r}) \cdot f(t)$. Then

$$\nabla^2[g(\mathbf{r}) \cdot f(t)] = \frac{1}{c^2} \frac{\partial^2}{\partial t^2}[g(\mathbf{r}) \cdot f(t)],$$

$$f(t)\nabla^2 g(\mathbf{r}) = \frac{g(\mathbf{r})}{c^2} \frac{\partial^2 f(t)}{\partial t^2},$$

$$\frac{\nabla^2 g(\mathbf{r})}{g(\mathbf{r})} = \frac{1}{c^2 f(t)} \frac{\partial^2 f(t)}{\partial t^2}.$$

The left and right side of this equation are only functions of $\mathbf{r}$ and t respectively, and can therefore be set to the same constant, say $-k^2$, which can be seen as the square of a vector, $k^2 = |\mathbf{k}|^2$. We obtain the two equations:

$$\nabla^2 g(\mathbf{r}) = -k^2 g(\mathbf{r}),$$

$$\frac{\partial^2 f(t)}{\partial t^2} = -c^2 k^2 f(t).$$

The first equation can be written:

$$\left(\frac{\partial^2}{\partial x^2} + \frac{\partial^2}{\partial y^2} + \frac{\partial^2}{\partial z^2}\right) g(x, y, z) = -(k_x^2 + k_y^2 + k_z^2) g(x, y, z),$$

and has the solution:

$$
\begin{aligned}
g(x, y, z) &= \exp[\pm i (k_x \cdot x + k_y \cdot y + k_z \cdot z)], \\
g(\mathbf{r}) &= \exp(\pm i\, \mathbf{k} \cdot \mathbf{r}).
\end{aligned}
$$

The other equation has the solution:

$$f(t) = \exp(\pm ickt),$$

By combining the two solutions, we obtain ($\psi = gf$):

$$
\begin{aligned}
\psi &= \exp(i\, \mathbf{k} \cdot \mathbf{r}) \cdot \exp(-ickt), \\
\psi &= \exp\left[i(\mathbf{k} \cdot \mathbf{r} - \omega t)\right].
\end{aligned}
\tag{7.8}
$$

This is an equation for a wave at a definite point in space, $\mathbf{r}$, which oscillates with a frequency $\omega = ck$. The frequency corresponds to the frequency of the string from which it stems, so ω is related to the total energy of the string, E, and we obtain:

$$
\begin{aligned}
E &= \omega \hbar, \\
\omega &= \frac{E}{\hbar}.
\end{aligned}
$$

The property k is a number which in quantum mechanics is termed the *wave number* $\mathbf{k}$. Its norm is related to the frequency and hence the energy of the system:

$$\|\mathbf{k}\| = k = \frac{\omega}{c} = \frac{E}{\hbar c},$$

which indicates that it is related to the momentum, $\mathbf{p}$, of the entire system given by (see 4.17):

$$\mathbf{k} = \frac{\mathbf{p}}{\hbar}.$$

With these properties inserted, (7.8) takes the form:

$$\psi = \exp\left[\tfrac{i}{\hbar}(\mathbf{p}\cdot\mathbf{r} - Et)\right].$$

Now, say that there in the same vicinity are two particles, or systems, say α and β, with the energies E_α and E_β, and momentums $\mathbf{p}_\alpha$ and $\mathbf{p}_\beta$ respectively. Both systems generate local displacements at $\mathbf{r}$, and these components can be added to a composite state $\psi_{\alpha\beta}$ given by:

$$\begin{aligned}
\psi_{\alpha\beta} &= \exp\left[\frac{i}{\hbar}\big((\mathbf{p}_\alpha + \mathbf{p}_\beta)\cdot\mathbf{r} - (E_\alpha + E_\beta)t\big)\right] \\
&= \psi_\alpha \cdot \psi_\beta.
\end{aligned} \tag{7.9}$$

We notice that the state stemming from a number of systems is just the product of the state components stemming from each system. However, we must not forget that we are moving about in a mathematical realm where we have no idea of where the energy really is situated. The only thing we know, is that the energy is somewhere around. The same applies to momentum. If a system of deformations moves along, it brings with it an amount of energy and it has momentum, but that is all we know (see Kelvin's theorem Sec. 3.7, page 45).

In this model a composite particle consists of a number of wave packets moving with the speed of light in some direction. Let us call any composite particle for a material body, or just a body, and let the wave packets be synonymous with elementary particles.

Now let us consider a body at a position so far from the origin that any points inside the body can be said to have the same radius vector $\mathbf{r}$, and let the body consist of m elementary particles each with energy E_n and momentum $\mathbf{p}_n$. Then the wave function for particle number n is given by:

$$\psi_n = \exp\left[\tfrac{i}{\hbar}(\mathbf{p_n}\cdot\mathbf{r} - E_n t)\right].$$

The wave function for the whole body is the product of all the wave functions inside the body, and we obtain:

$$\psi_b = \psi_1 \cdot \psi_2 \cdot \psi_3 \cdots \psi_m$$

$$= \exp\left[\frac{i}{\hbar}\left(\sum_{n=1}^{m} \mathbf{p}_n \cdot \mathbf{r} - \sum_{n=1}^{m} E_n t\right)\right]$$

$$= \exp\left[\frac{i}{\hbar}(\mathbf{p_b} \cdot \mathbf{r} - E_b t)\right].$$

The elementary particles inside the body are bound together by almost unbreakable strings tying them together, so for every wave packets moving in one direction, there is another packet moving in the opposite direction keeping the momentums in balance. Nevertheless each doublet, triplet, or whatever, may have a net momentum in the $\mathbf{v}$-direction, and the adds-up of all the momentums in the system is the body's net momentum $\mathbf{p}_b$.

Recall that the mathematics of deformations in the spatial continuum belongs to a Lorenz group, and transforms as such (see Chap. 4). So first I will consider a body at rest in a primed coordinate system. Then the net momentum is nil, and in this coordinate system the wave function reduces to:

$$\psi_b' = \exp\left[-\frac{i}{\hbar}E_b t'\right].$$

Next I will transform the function over to an unprimed coordinate system moving with speed $-\mathbf{v}$. In the unprimed frame of an observer, the body is moving in the positive $\mathbf{v}$ direction. It is only by transformations between Lorenz coordinate systems that the divergence-free properties are invariant[5], so we have got to apply the inverse Lorenz transformations:

$$\mathbf{r}' = \frac{\mathbf{r}+\mathbf{v}t}{\sqrt{1-v^2/c^2}},$$

$$t' = \frac{t+\mathbf{v}\cdot\mathbf{r}/c^2}{\sqrt{1-v^2/c^2}}.$$

We obtain:

$$\psi_b = \exp\left[\left(-\frac{i}{\hbar}E_b\right)\cdot\left(\frac{t-\mathbf{v}\cdot\mathbf{r}/c^2}{\sqrt{1-v^2/c^2}}\right)\right].$$

---

[5]Discussed in Chap. 4.

The spatial condition in which the body resides may well vary as a function of position and time. That might result in a change of the body's internal energy. In order to take this possibility into account, I will take the energy at a given position and time to be E_0. To this initial energy I will add the change given by $V(\mathbf{r}, t)$ such that the internal energy of the body is given by $E_b = E_0 + V$. We obtain:

$$\psi_b = \exp\left[\left(-\tfrac{i}{\hbar}(E_0 + V)\right)\left(\frac{t - \mathbf{v}\cdot\mathbf{r}/c^2}{\sqrt{1 - v^2/c^2}}\right)\right].$$

Then the wave function for the body takes the form:

$$\psi_b = \exp\left[\left(-\frac{i}{\hbar}(E_0 + V)t + \frac{i}{\hbar}(E_0 + V)\frac{\mathbf{v}\cdot\mathbf{r}}{c^2}\right)\frac{1}{\sqrt{1 - \frac{v^2}{c^2}}}\right]$$

$$\approx \exp\left[\frac{i}{\hbar}\left(\frac{E_0}{c^2}(\mathbf{v}\cdot\mathbf{r}) + \frac{V}{c^2}(\mathbf{v}\cdot\mathbf{r}) - E_0 t - Vt\right)(1 + \frac{v^2}{2c^2})\right]$$

$$= \exp\left[\frac{i}{\hbar}\left(\frac{E_0}{c^2}(\mathbf{v}\cdot\mathbf{r}) + \frac{V}{c^2}(\mathbf{v}\cdot\mathbf{r}) - E_0 t - Vt\right.\right.$$
$$\left.\left. + \frac{v^2}{2c^2}\frac{E_0}{c^2}(\mathbf{v}\cdot\mathbf{r}) + \frac{v^2}{2c^2}\frac{V}{c^2}(\mathbf{v}\cdot\mathbf{r}) - \frac{v^2}{2c^2}E_0 t - \frac{v^2}{2c^2}Vt\right)\right].$$

Notice that at this point I have made an approximation such that the rest of the development is restricted to small velocities v compared to c.

The wave function ψ_b has several overtones, and here I will single out the wave function with the highest frequency and longest wavelength (highest wave number), which I assume is representing the unobservable internal oscillations, ψ_i, while ψ is the observable wave equation ($\psi_b = \psi_i \cdot \psi$). We get the two wave equations:

$$\psi_i = \exp\left[\frac{i}{\hbar}\left(\frac{v^2}{2c^4}E_0(\mathbf{v}\cdot\mathbf{r}) - E_0 t\right)\right],$$

$$\psi = \exp\left[\frac{i}{\hbar}\left(\frac{1}{c^2}(E_0 + V + \frac{v^2}{2c^2}V)(\mathbf{v}\cdot\mathbf{r}) - (\frac{v^2}{2c^2}E_0 + V + \frac{v^2}{2c^2}V)t\right)\right].$$

Since $V \ll E_0$ the only term with V that may influence the wave function in any significant degree is Vt. Hence the other terms may be skipped.

$$\psi = \exp\left[\frac{i}{\hbar}\left(\frac{1}{c^2}E_0(\mathbf{v}\cdot\mathbf{r}) - \left(\frac{v^2}{2c^2}E_0 + V\right)t\right)\right].$$

Let us take the Laplacian and the time derivative of ψ and see where that leads us.

$$\nabla^2\psi = \frac{i^2 v^2 E_0{}^2}{\hbar^2 c^4}\psi_2$$

$$\frac{-\hbar^2 c^4}{v^2 E_0{}^2}\nabla^2\psi = \psi,$$

and

$$\frac{\partial\psi}{\partial t} = -\frac{iv^2 E_0}{2c^2\hbar}\psi - \frac{iV}{\hbar}\psi$$

$$-\frac{2c^2\hbar}{iv^2 E_0}\frac{\partial\psi}{\partial t} - \frac{2c^2 V}{v^2 E_0}\psi = \psi$$

We have two expressions that both are like ψ, so we can draw them together:

$$\frac{-\hbar^2 c^4}{v^2 E_0{}^2}\nabla^2\psi = -\frac{2c^2\hbar}{iv^2 E_0}\frac{\partial\psi}{\partial t} - \frac{2c^2 V}{v^2 E_0}\psi,$$

$$\frac{-\hbar^2 c^2}{2E_0}\nabla^2\psi = -\frac{\hbar}{i}\frac{\partial\psi}{\partial t} - V\psi.$$

We finally obtain the equation:

$$\frac{-\hbar^2}{2m}\nabla^2\psi + V\psi = i\hbar\frac{\partial\psi}{\partial t}$$

which is the the famous Schrödinger wave equation, and like the Schrödinger equation, it is only valid for small velocities when $v \ll c$. It has played an important role in the development of quantum mechanics since its discovery by Erwin Schrödinger (1887-1961) in the 1920th.

Part IV

Gravitation

Chapter 8

VLS theory of gravity

And God said, "Let there be a dome in the midst of the waters, and let it separate the waters from the waters." (Genesis 1.6)

Confined wave energy will displace some amount of spatial mass almost like a bubble of gas in a compressed liquid. When the pressure drops, the liquid will expand, but the bubble will expand much more in order to adapt to the lower pressure. The pressure in the bubble and the receding surface causes the confined kinetic energy to give away some of its energy, which has got to be carried away as a non harmonic P-wave. The P-wave takes the form of a gradually decreasing pressure and mass density away from the confined wave energy. Hence the wave speed around a body with confined energy in expanding space will increase with distance and all waves in a test body will be deflected towards the other body. This is a variable light speed (VLS) theory of gravity, which at first was proposed by Einstein in 1911 but later abandoned. In this model it is essential.

8.1 Confined energy in expanding space

A material body in this model is basically confined wave energy moving around inside the boundary of the body in some irregular ways, but always with the speed of light. Elastodynamic waves, like all kinds of wave movements, can be reflected, undergo refraction, and interfere in both constructive and destructive manners. The energy in a body is huge, so it is convenient to introduce another property, *mass*, defined as the energy divided by the speed of light squared.

The energy distribution within a body is normally not smooth, but concentrated in small energy packets that again may be divided into still smaller packets and so on down to the smallest possible sizes, which we can call elementary particles. By adding up the mass inside a small area that still is great in relation to the size of the elementary particles and dividing it by the volume, we obtain the mean density, or more specific, the *mass density* of the body.

Confined disturbance energy will exert a pressure on whatever keeps it at bay. It is irrelevant in this connection to focus on what keeps it confined, but it is important to notice that radiation energy always will introduce a pressure into the spatial continuum where it is confined. The force on an enclosing surface is uniformly directed in all directions and is equal to 1/3 of the confined energy density. This is a general principle in physics. From [18] I quote the very precise statement:

> It may be shown by electromagnetic theory, by quantum theory, or by thermodynamics, making no assumptions as to the nature of the radiation (other than isotropy), that the pressure against a surface exposed in a space traversed by radiation uniformly in all directions is equal to one third of the total radiant energy per unit volume within that space.

Normally this force is rather weak, but with a high energy density, as we find it in a material body, it may all the same be significant.

Confined energy in a static, not expanding, spatial continuum will displace some spatial mass, but it can be read directly out of the elastodynamic equation, that even if the spatial continuum outside the energy concentration is somewhat deformed with diverging stress components in different spatial directions, the mean stress is constant throughout the whole space, and the spatial mass density is uniform. Just think of a gas bubble in a liquid. There will be no pressure gradient around it, and the pressure in the gas bubble and the liquid are the same. However, this situation is changed if the energy concentration is located in an expanding space. Then the energy pressure inside will exert a force on a receding surface in the expanding spatial continuum, and the system will loose energy. It is important to notice that the energy loss is independent of the size of the enclosure (10.3). Therefore it does not matter how the energy is distributed within a body; the energy loss depends only on the confined energy and the expansion rate of the space around it.

As discussed in Sec. 1.2 the present space is expanding, not only on a cosmic scale, but all the way down to the smallest elementary particles. The rate of expansion is governed by the Hubble law, which says that two points in space move away from each other with a speed given by the distance between the points times Hubble's constant. The Hubble 'constant' is supposed to be constant, but may be linked to the inverse of the age of the Universe, which is a very small number that is almost constant in the course of the time span of man's existence on Earth. Hence it can be considered constant for all practical purposes.

8.2   The gravitational potential

The body force that confined energy exerts in the spatial continuum, can be inserted into the Navier-Cauchy equation, and - as mentioned above - we find that in a static, none expanding spatial continuum, the pressure outside the confined energy is constant everywhere. Not so in an expanding spatial continuum. As a result of delay, the pressure in the confined energy takes some time to adapt to the declining pressure farther away, and we get a retarded pressure that declines as an inverse function of the distance from the body.

To get an idea of what is going on we can think of a gas bubble immersed in a liquid under pressure. The fluid is almost incompressible, but not quite, so it will expand a little as the pressure falls. The gas in the bubble, however, is highly compressible and will expand significantly more than the fluid, hence its diameter will grow. The pressure in the gas exerts a force on the inner wall of the bubble and the work done per unit time is like the change in diameter times the pressure times the surface divided by the change in time. The work done by the confined energy has got to be taken from the kinetic energy in the bubble, and it somehow has got to be carried away. In an adiabatic process the only feasible means is by some kind of wave movement, but it cannot for obvious reasons be by a harmonic wave.

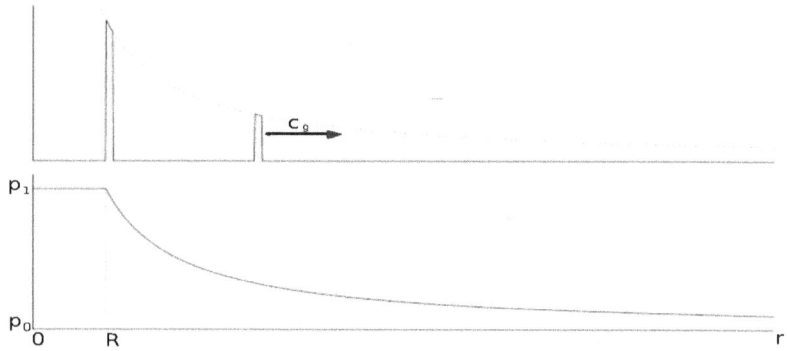

Figure 8.1: The pressure field around a gas bubble with radius R in a liquid.
Upper graph: *A short raise in the pressure inside the bubble generates a pressure pulse moving outwards from the bubble with the speed c_g, while the pressure amplitude falls off inversely proportional to the distance from the center of the bubble*
Lower graph: *If the pressure in the bubble, p_1, is forced to remain on a slightly higher level than the liquid pressure, p_0, then the pressure around the bubble will fall off to the liquid pressure inversely proportional to the distance from the center, precisely like the pressure in the pulse above.*

Say that we somehow can control the pressure inside the bubble. First we raise the gas pressure slightly above the liquid pressure for a short time (see upper graph in Figure 8.1). It will first compress the liquid near the surface of the bubble, which then spread as a compression pulse in the liquid with the speed of a P-wave. The amplitude will decrease towards the fluid pressure inversely proportional to the distance from the bubble. Next we let the gas pressure remain at the somewhat elevated pressure over a long time. Like with a single pulse, the overpressure will fall off away from the bubble inversely proportional with the distance from the center (see lower graph in Figure 8.1).

It is difficult to see how we could control the pressure in the bubble, but if the pressure in the liquid instead should fall because of expansion, the situation would be much like the same. It will be more like a shock wave from an ongoing underwater explosion, but quite a less dramatic than that. For the time it takes for the overpressure to fall down to zero, it will be an ongoing process. The only way to figure what the wave might be like, is by thinking that the decreasing pressure in the bubble is delayed a little in relation to the all over pressure in the fluid far away from the bubble. The pressure in the fluid right outside the bubble, however, has got to be like that in the gas inside and gradually decreasing with distance down to the pressure far away. This is a dynamic process and the energy is carried away as an anharmonic wave spreading away in all directions. Like any spherical wave, its amplitude, i.e. pressure, will fall off inversely proportional to the distance. Thus we find a pressure gradient around the bubble, and with a pressure gradient the spatial density varies, and so does the speed of say sound waves. It is increasing with distance from the confined energy alongside with the decreasing density.

This is about what has got to happen with confined energy in expanding space. Around a body with confined energy in expanding space there will be a density gradient. (I'll return to the question in Chapter 10 by solving the Navier-Cauchy equation.) Like a gas bubble in a fluid with constant pressure, confined energy in a non expanding space does not create any density gradient around itself. It is this density gradient that can be identified as the gravitational potential. It will relate to the amount of confined energy, the mass density, and to the Hubble constant. As far as we know, the Universe has expanded since its birth in the Big Bang some 13.8 billion years ago, but on the scale of man's existence on Earth, the degree of expansion is minute and the rate of expansion can be considered constant. The combined effect of confined energy and the expanding universe should also be extremely small, and — compared with other forces — indeed it is. We have got to remember what enormous amount of energy there is in one gram of matter. Einstein gave us the formula, $e = mc^2$. Nevertheless the gravitational attraction between two atoms is not detectable. In fact gravitational forces are almost 40 orders of magnitude smaller that electromagnetic forces. On a greater scale, however, other forces of nature are absent and gravitation becomes dominant.

This is just a qualitative discussion on how the speed of light varies with distance, but a more thorough discussion has got to be postponed to Chapter 10. Thus, in this model the speed of light is not constant as commonly regarded, but depends on the position in a gravitational field. It is slower near a gravitating body than farther out. The classic property, the gravitational potential, is here linked to the spatial density. Variable Light Speed (VLS) theories have popped up repeatedly during the 20th century. In fact Einstein himself started his work on gravity by proposing that the speed of light varies in a gravitational field, but he discarded it after a couple of years and came up with his principle of general relativity in 1916. But this model of space and matter hinges on the conception that the speed of light is not constant. This calls for some explanation, since all measurements show that the light speed seems to be a universal constant, which of cause is proven beyond any doubt. Here I shall only state that at least our time measuring devises are shown to slow down when getting deeper into a gravitational field [7], and if we base our standard length unit on the wavelength of radiation from our reference oscillator, we would measure the same light speed regardless of whatever the real light speed might be. For example if the length unit is defined as one wavelength of the emission from a given element and the time unit is defined as one period of the emission, then the speed of light would be measured to one regardless of what the the real speed of light might be. We would need a demon who could define a length and time unit far away from any gravitating object, and bring the units with him to wherever he wants to measure the real light speed. The light speed measured by him would be a number slightly less than one near a gravitating object[1]. But unfortunately we do not have access to his units and thus are doomed to measure the same light speed everywhere.

Finally, since the young Universe was denser than today's, the light from far away galaxies should be more red-shifted than the expansion should indicate and we could come to believe that expansion is accelerating. Without further discussion we also notice that if the spatial continuum instead of expanding should be contracting, then all processes would be reversed and we would have a negative gravitation.

8.3 Matter in inhomogeneous space

We have seen that a material body creates around itself a field with increasing wave speed away from the body. The next goal is to figure out how a test body would behave in such a field. In this model of space and matter, material bodies are nothing but confined waves energy moving with the speed of light. Basically the waves take the form of spherical waves that are moving towards singularities from where they are reflected and redirected towards other singularities in an eternal round-dance inside the body. Waves passing through an environment with varying wave speeds will always be deflected towards areas with higher densities and lower wave speeds, and thus all waves inside a test body will continually be deflected towards the heavier gravitating body. Hence the test body will get an acceleration in that direction.

To get an impression on what is going on, we can compare a

---

[1]In some connections it may be necessary to make a clear distinction between the measured and the real speed of light. Then c is meant to be the universal constant and c_l the real (variable) speed of light.

wave-front with a troop of soldiers marching in step with each other (see Figure 8.2). Like the soldiers the wave-front is moving towards the focal point where it is reflected. If the wave speed is uniform the wave front will form concentric circles as it moves inwards, but if there is a density gradient as it will be around a gravitating body, the wave front continually will be deflected towards the body both when it travels inwards as well as after it has been reflected.

This happens to all wave movements within the test body, say an object on Earth, and the whole body gets into a state of acceleration towards the center of the Earth. From this angle of view, gravitation is not a force, but a property of the spatial continuum itself that causes the acceleration. All matter in the vicinity gets the same acceleration independent of its mass. Gravitational forces only occur when the body is hindered from moving freely in space. A more direct consequence of this effect is the bending of light from faraway stars when it passes near by the sun. This was first seen during the solar eclipse on May 29, 1919. It was then taken as a proof of Albert Einstein's relativity theory, but it fits equally well into this model.

Because of the slower wave movements near a heavy star, all processes there evolves slower than in intergalactic space, and the frequency of waves emitted from any exited atom is red shifted. Astronomers see this effect when they study spectra from stars they observe, and as we use this effect when we make our atomic clocks, they will run slower on a heavy planet than on a lighter one, and slower on Earth than in interplanetary space. If our length units are defined as wavelengths of light with certain frequencies, we must expect that even our length units wary accordingly. So, since our measuring units change depending on position, we need a new way to cope with measurements and interpretations of what we are seeing around us. That is the tool that General Relativity (GR) provides us with. Instead of varying clocks and meter sticks, space itself is considered to be curled.

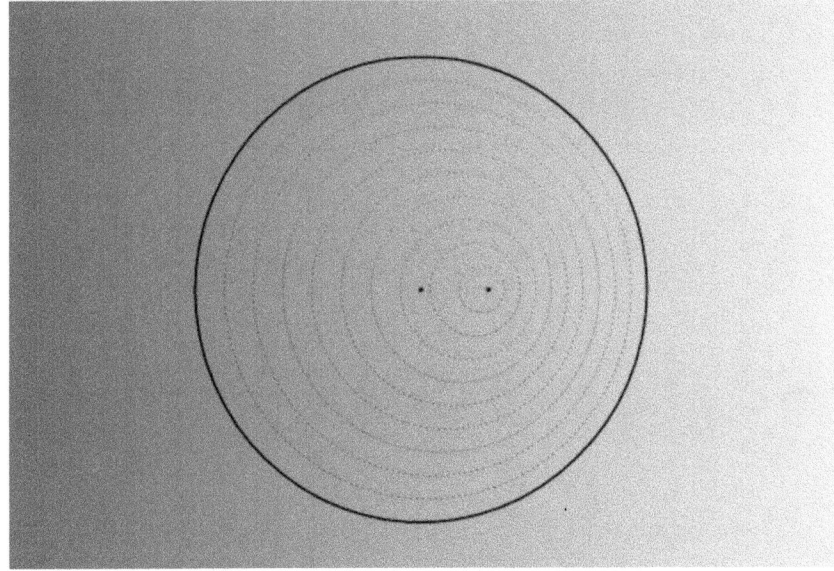

Figure 8.2: *Let a troop of soldiers be lined up along the circumference of a circle, face inwards. If they march on a level ground with
equally long steps, they will meet at the center. If they march on
a slop tilting to the right, however, the rightmost soldiers have got
to march uphill, while the soldiers to the left are marching downhill. The soldiers marching downhill are supposed to take slightly
longer steps than those who are marching uphill, and those who are
coming in from the flanks, are supposed to take slightly longer steps
with their uphill legs than the other ones. The result will be that the
whole troop is falling off in the downward direction and are going
to meet at a point somewhat downhill in relation to the center of
where they started.*

Chapter 9

The expanding space

In this model space is a compressed elastic continuum of infinite extension or nearly so. It can be treated by the linear theory of elasticity, and in its present state of compression it has a given mass density and an inertia that is proportional to the density. How it gets its inertia is uncertain, but here it will be treated as a given property that increases with increasing spatial density. If space is utterly compressed from this state, the density increases and the wave speed is slowing down. This property gives room for a variable wave speed as a function of position. Since the creation by the Big Bang, the spatial continuum has expanded by a factor given by Hubble's constant, which may or may not be a real constant.

9.1 The compressed space

Let space be a homogeneous and isotropic manifold of almost infinite extension, which initially was compressed to some extent and at present is expanding uniformly in all directions. Let it also be an elastic continuum that for small deformations can be described by *the linear theory of elasticity*. The equation of deformation and motion can be written as the Navier-Cauchy equation (3.7):

$$c_g{}^2 \operatorname{grad} \operatorname{div} \mathbf{u} - c_l{}^2 \operatorname{curl} \operatorname{curl} \mathbf{u} + \frac{\mathbf{b}}{\rho_s} = \ddot{\mathbf{u}}.$$

The property **b** is a body force from the outside world, e.g. gravitation, but if confined disturbance energy is present, it may act upon the spatial continuum with a resulting force that can be represented by **b**. Solenoidal waves travel with the speed:

$$c_l \ = \ \sqrt{\tfrac{\mu_s}{\rho_s}}, \tag{9.1}$$

and irrotational waves with the speed

$$c_g \ = \ \sqrt{\tfrac{\lambda_s + 2\mu_s}{\rho_s}}. \tag{9.2}$$

I put the index s (for *spatial*) on ρ, μ and λ in order to distinguish them from other uses of the same Greek letters. For example ρ_m is the density of matter while ρ_s is the density of the spatial continuum.

If the spatial continuum is locally deformed, the deformational energy is given by an integral over the deformed area (3.13):

$$E = \frac{1}{2} \int_B \left[\rho_s \dot{\mathbf{u}}^2 + (\lambda_s + 2\mu_s)(\operatorname{div} \mathbf{u})^2 + \mu_s(\operatorname{curl} \mathbf{u})^2 \right] dv.$$

A formula of this type can be interpreted 'as if' the energy density is given by:

$$e = \frac{1}{2} \rho_s \dot{\mathbf{u}}^2 + \frac{1}{2} (\lambda_s + 2\mu_s)(\operatorname{div} \mathbf{u})^2 + \frac{1}{2} \mu_s(\operatorname{curl} \mathbf{u})^2,$$

but then one must have in mind its limitation. The energy can in reality be found anywhere.

If there is a certain degree of uniform compression in the spatial continuum, that also will represent energy. When I deduced the energy equation above from the Navier-Cauchy equation in Sec. 3.7, however, I also found that the energy is independent of any possible initial uniform pressure. Hence such energy will not affect our calculations. In reality it is a hidden energy.

9.2 The expanding space

In this model space consists of a homogeneous and isotropic elastic continuum filling up the whole Universe. The Universe was created in the Big Bang that probably was preceded by a Big Crunch – an inwards moving spherical compression wave that ended in a boiling inferno where all matter was created. From then on the compressed spatial continuum has expanded in all directions in a fairly orderly manner behind a frontal wave, similar to a tidal bore, that divides the pressure on the inside from the much lower pressure on the outside.

Some fraction of a second after the Big Bang, the whole universe was compressed into a very small volume, and the distance between any two points was negligible in relation to what the distance is today. Consider two points at a present distance, a, from each other. Perhaps we by observation can find out that they are moving away from each other with the speed v. If the separating speed has been constant during the whole lifetime of the universe, we find that the elapsed time since the points were close together is:

$$T = \frac{a}{v},$$

and so we can set up the relation:

$$v = \frac{1}{T}a.$$

If T is extremely large, we can replace $1/T$ with a property, H, that does not change much in the course of a short time interval of observation, and hence can be approximated as being constant. Then we have a formula that tells us the speed at which two arbitrary points in space are moving away from each other:

$$v = Ha,$$

and:

$$H = \tfrac{1}{T}. \tag{9.3}$$

The Hubble Space Telescope Key Project Team has measured Hubble's constant with an uncertainty of 10 percent to:

$$H = 70 \text{ km s}^{-1}\text{Mpc}^{-1} = 2.3 \times 10^{-18}\text{sec}^{-1}.$$

If we use this property on the equation above we obtain:

$$
\begin{aligned}
T &= \frac{1}{2.3 \cdot 10^{-18}} \sec = \frac{10^{18}}{2.3 \cdot 3600 \cdot 24 \cdot 365} \text{ year} \\
&= 13.8 \cdot 10^{9} \text{ year,}
\end{aligned}
$$

which is in excellent agreement with the estimated age of the Universe. It might be a coincidence, but we can not exclude the possibility that the Hubble constant is a function of the age of the Universe either.

9.3 Variable speed of light

In order to convey shear and compression waves, the spatial continuum has got to have a mass density, which can be set to ρ_s, but it is hard to say how space has got its inertia. I suspect that it originates from some of the residual energy in the Big Crunch, but here I shall only assume that the spatial continuum in intergalactic space at present has a certain intrinsic density, ρ_{s0}, that increases with compression. The mass inside a volume is not supposed to change if it is compressed, hence the density increases:

$$
\begin{aligned}
V_0 \cdot \rho_{s0} &= V \cdot \rho_s = m, \\
\rho_s &= \rho_{s0}\frac{V_0}{V_0 + \Delta V} \\
&= \rho_{s0}\frac{1}{1 + \frac{\Delta V}{V_0}} \\
&= \rho_{s0}\frac{1}{1 + \operatorname{div} \mathbf{u}}.
\end{aligned}
\tag{9.4}
$$

Finally Lamé's parameters, μ and λ, are assumed to be constant for small deformation changes in the present state of the spatial continuum:

$$\mu_s = \mu_{s0}.$$

If the spatial density varies throughout space, then so does the speed of light. We have:

$$
\begin{aligned}
c^2 &= \frac{\mu_s}{\rho_s} \\
&= \frac{\mu_{s0}}{\rho_{s0}}(1 + \operatorname{div} \mathbf{u}) \\
&= c_l^2(1 + \operatorname{div} \mathbf{u}). \tag{9.5}
\end{aligned}
$$

Here c_l is the speed of light at infinity, and c the local speed of light.

Chapter 10

Newtonian gravitation

Newton discovered the law that two heavenly bodies attract each other with a force proportional to the product of their masses divided by the distance between them squared. This is a force that acts at a distance. A more modern view is to consider gravitation as an effect of a gravitational potential, the gradient of which gives the gravitational acceleration at any point in space. A still more modern view is to apply the General Theory of Relativity, GR, where bodies move along geodesics in curved space-time. In this chapter the gravitational potential is shown to be a function of the compression of the spatial continuum. Its strength is derived from the amount of confined energy and the expansion of space. A test body consisting of a small amount of confined wave energy, will get an acceleration towards the confined energy in agreement with what is predicted by Newtons gravitational law.

10.1 Pressure from matter

Of course no physical objects can ever pass through an elastic continuum, so matter in this model has got to be nothing but confined disturbance energy. Since the N-C equation predicts that disturbance waves can be of two mayor types, longitudinal (curl-free) and transversal (divergence-free) waves, it will be necessary to decide what type of waves matter consists of. We already know that electromagnetism constitutes a mayor part of the binding forces that bind particles together, hence it is natural to assume that matter basically is an electromagnetic phenomenon. Here I shall stick to that conception without any further comments. Electromagnetic energy and momentums can unambiguously be described by the stress energy tensor:

$$T^{\alpha\beta} = \begin{bmatrix} e & S_x/c_l & S_y/c_l & S_z/c_l \\ S_x/c_l & -\sigma_{xx} & -\sigma_{xy} & -\sigma_{xz} \\ S_y/c_l & -\sigma_{yx} & -\sigma_{yy} & -\sigma_{yz} \\ S_z/c_l & -\sigma_{zx} & -\sigma_{zy} & -\sigma_{zz} \end{bmatrix},$$

where e is the energy density, $\mathbf{S}$ the energy flow vector, σ Maxwell's stress tensor, and c_l the speed of light.

The trace of the tensor is found to be vanishing, hence[1]:

$$\operatorname{tr} \mathbf{T} = T^{\alpha}{}_{\alpha} = -e - \sigma_{xx} - \sigma_{yy} - \sigma_{zz} = 0.$$

The energy in a body with confined energy can be thought of as divided into small wave packets, which at a given time are moving in all possible directions such that their mean velocity is zero, or else if the body is moving, the mean velocity is like that of the body itself. The mathematical identity, $\operatorname{tr} A + \operatorname{tr} B = \operatorname{tr}(A + B)$, tells us that we can add up the different wave packets to encompass the trace of the whole body.

If the normal stress components are negative, the forces are directed inwards into a volume element and the radiation pressure is positive. Hence the normal stress in any spatial direction is

$$\sigma_{xx} = \sigma_{yy} = \sigma_{zz} = -p,$$

and the equation above yields[2]:

$$p = \frac{1}{3}e. \tag{10.1}$$

---

[1] Notice that e changes sign when one of the indices in $\mathbf{T}$ is lowered.

[2] By the way, this applies to to all kinds of confined radiation. It exerts a pressure on whatever keeps it at bay like one third of the energy density.

Generally we haver $\mathbf{b} = \nabla \cdot \sigma$, and we obtain

$$b_j = \sigma_{ii,j} = -p_{,j}$$
$$\mathbf{b} = -\mathrm{grad}\,p. \qquad (10.2)$$

However small volumes of radiation we consider within certain limits, the mean energy distribution is assumed to be uniformly moving in all directions, i.e. isotropic, so the formula above is supposed to be valid for all energy within a material body.

10.2 Energy loss

In an expanding space, a surface around a sphere with radius, r, will recede from the origin with a speed, $v = H \cdot r$. The confined energy within the sphere will loose energy at a pace given by the force acting on the surface, $a = 4\pi r^2$, times the velocity, v. The pressure on the surface is given by $p = 1/3\,e$, and the energy lost per unit time becomes:

$$\frac{\partial E}{\partial t} = -a \cdot p \cdot v$$
$$= -4\pi r^2 \cdot \frac{1}{3}e \cdot Hr,$$
$$\frac{1}{4/3\,\pi r^3}\frac{\partial E}{\partial t} = -He,$$

so:

$$\frac{\partial e}{\partial t} = -He. \qquad (10.3)$$

Confined energy will loose energy in the expanding space, but since all occurrence of energy is covered by the same stress energy tensor, even free energy will loose energy at the same rate. Cosmic microwave background radiation has continually lost energy since it was created by thermal radiation in the early universe.

We can differentiate e once more with respect on time, and on the assumption that $H = 1/T$, we obtain:

$$\frac{\partial^2 e}{\partial t^2} = -\frac{\partial e}{\partial t} H - e \frac{\partial H}{\partial t}$$

$$= -(-eH)H - e\left(-\frac{1}{T^2}\right),$$

$$\frac{\partial^2 e}{\partial t^2} = 2eH^2. \tag{10.4}$$

If H is constant, we get the same expression without the factor of two. It will not make any significant difference in later deductions.

Notice that although the fraction of energy loss is extremely small, the energy in a material body is so huge that the energy loss cannot be ignored.

10.3   The gravitational potential

We have seen that confined isotropic radiation will act as a body force which can be inserted into the Navier-Cauchy equation (3.7) in order to find the deformation it generates in the spatial continuum. By (10.2) we obtain:

$$c_g{}^2 \text{grad div} \mathbf{u} - c_l{}^2 \text{curl curl } \mathbf{u} - \frac{\text{grad } p}{\rho_s} = \ddot{\mathbf{u}}.$$

By applying the divergence operator on the equation we obtain:

$$\nabla^2(\text{div } \mathbf{u}) - \frac{1}{c_g{}^2}\frac{\partial^2(\text{div } \mathbf{u})}{\partial t^2} = \frac{\nabla^2 p}{\rho_s c_g{}^2},$$

and next adding the same term:

$$\frac{\partial^2}{\partial t^2}\left(\frac{p}{\rho_s c_g{}^2}\right),$$

to both sides of the equation and ordering, we obtain

$$\nabla^2\left(\text{div } \mathbf{u} - \frac{p}{\rho_s c_g{}^2}\right) - \frac{1}{c_g{}^2}\frac{\partial^2}{\partial t^2}\left(\text{div } \mathbf{u} - \frac{p}{\rho_s c_g{}^2}\right) = \frac{\partial^2}{\partial t^2}\left(\frac{p}{\rho_s c_g{}^4}\right).$$

By (10.1) and a new scalar, Φ, defined as:

$$\Phi = -\frac{c_l^2}{2}\,\mathrm{div}\,\mathbf{u},\tag{10.5}$$

the equation above takes the form

$$\nabla^2\left(\Phi + \frac{pc_l^2}{2\rho_s c_g^2}\right) - \frac{1}{c_g^2}\frac{\partial^2}{\partial t^2}\left(\Phi + \frac{pc_l^2}{2\rho_s c_g^2}\right) = \frac{c_l^2}{6\rho_s c_g^4}\frac{\partial^2 e}{\partial t^2}.$$

By (10.4) and the two new definitions:

$$G = \frac{H^2}{12\pi\rho_s}\frac{c_l^4}{c_g^4},\tag{10.6}$$

$$\rho_m = \frac{e}{c_l^2},\tag{10.7}$$

we finally obtain

$$\nabla^2\left(\Phi + \frac{pc_l^2}{2\rho_s c_g^2}\right) - \frac{1}{c_g^2}\frac{\partial^2}{\partial t^2}\left(\Phi + \frac{pc_l^2}{2\rho_s c_g^2}\right) = 4\pi G\rho_m.$$

The three defined properties are the *gravitational potential*, Φ, the *gravitational constant*, G, and the *mass density of matter*, ρ_m.

Outside the material body the pressure from matter is zero and we obtain

$$\nabla^2\Phi - \frac{1}{c_g^2}\frac{\partial^2\Phi}{\partial t^2} = 4\pi G\rho_m\tag{10.8}$$

This equation for the gravitational potential was first formulated by Gunnar Nordstrom (1881-1923) in 1912, except from the different wave speed, c_g, but a year later he dropped it and presented a slightly different equation. His first attempt clearly predicted a loss of energy from a gravitating system, so the new formulation seems to be an attempt to correct what nobody at the time were prepared to accept – the expanding universe.

Notice that in order to develop the equation for the gravitational potential above, I have assumed that the stresses in all directions are isotropic, which clearly is not the case for quickly moving bodies. Hence a complete theory has got to take such and other relations into consideration.

Provided that the scalar potential satisfies 'sensible boundary conditions at infinity and is consistent with causality' it has the general solution (see e.g. [4], Eq. 158):

$$\Phi(\mathbf{r}, t) = -G \int \frac{\rho_m(r', t - |r - r'|/c_g)}{|\mathbf{r} - \mathbf{r}'|} \mathrm{d}^3 r'.$$

It corresponds to the solution of Maxwell's equation for the scalar potential with the difference that the speed of light is exchanged for the speed of compressional waves. It can be interpreted as the potential at position r generated at a position, r', at an earlier time, $t - |r - r'|/c_{g_g}$, taken all over space, hence it is called a retarded potential. The variation of the strength of the source, however, is almost infinitesimal, except from a possibly detectable movement of the source point, r', so with this reservation in mind, the time delay can be neglected and the solution can be reduced to:

$$\Phi(\mathbf{r}, t) = -G \int \frac{\rho_m(r')}{|\mathbf{r} - \mathbf{r}'|} \mathrm{d}^3 r'.$$

This is the solution of an equation of the form:

$$\nabla^2 \Phi = 4\pi G \rho_m \tag{10.9}$$

which is the Poison's equation for the gravitational potential (see e.g. [5], Eq. 228).

As we have seen, it will be a very good approximation provided that the sources of the waves hardly have changed at all in the time span form the waves were emitted till their amplitudes were superposed at the observation point.

Equation 10.9 is an exact match to the classical equation for the gravitational potential. We notice that G can be interpreted as the gravitational constant[3], Φ as the gravitational potential, and

---

[3]In fact in this representation G is not a constant, but a function of H. Both are, however, functions of T^{-1} where T is supposed to be very great, hence the variation of G in the course of any time of observation is extremely small.

ρ_m as the mass density of the confined energy. Notice, however, that in this model it is (10.8) that is the correct equation describing the gravitational potential. The gravitational potential at the earth from the sun, for example, is the retarded potential as it were on the sun less than 5-6 minutes ago depending on the relation between between λ_s and μ_s. Like matter everywhere in the universe, the sun looses energy and mass all the time, but the loss in this short time interval is completely negligible, so on a planetary scale gravitation is not affected by the time delay. By greater distances, however, at a cosmological scale, it might be significant, but this will be needing a good deal of rethinking.

The total mass of a body is

$$M = \int_v \rho_m dV,$$

and if it is concentrated in a central symmetrical way around the point $r = 0$, then the potential at a distance r from that point is given by:

$$\Phi = -\frac{GM}{r} \qquad (10.10)$$

By taking the gradient of Φ we obtain:

$$\text{grad } \Phi = \frac{GM}{r^2}\hat{\mathbf{r}} \qquad (10.11)$$

10.4   Newton's gravitational law

We have seen that confined energy in an expanding space creates a compression field in its surroundings. The next goal becomes to find how a test mass behaves in such a field. First we recall that a material body basically consists of wave energy moving with the speed of light inside the body. The wave movements can be divided into two components: An internal component that adds up to zero, and a component that adds up to a velocity in some direction.

Equation 9.5 shows how the speed of light varies as a function of compression, and the compression field around a body is given

by the potential, (10.5). From here and onwards I term the local
speed of light just as $c = c(\mathbf{r})$ while c_l is the speed in intergalactic
space. We obtain:

$$c^2 = c_l{}^2 \left(1 - \frac{2}{c_l{}^2}\Phi\right),$$

and by taking the gradient, we obtain:

$$c \operatorname{grad} c = -\operatorname{grad}\Phi.$$

If, $2\Phi/c_l{}^2 \ll 1$, then the speed of light becomes:

$$c \approx c_l \left(1 - \frac{\Phi}{c_l{}^2}\right). \tag{10.12}$$

In a case where matter is organized in a central symmetrical manner
around an origin, then according to (10.11) it takes the form:

$$c \operatorname{grad} c = -\frac{GM}{r^2}\hat{\mathbf{r}}. \tag{10.13}$$

Consider a stationary flow of energy within the boundary of a
small test body. Divide the energy into small parts, e_n, that moves
along with the speed of light, c, in some direction, $\hat{\mathbf{p}}_n$. Then the
momentum of the packet is $\mathbf{p} = e_n/c \cdot \hat{\mathbf{p}}_n$. Since the total energy
has got to stay within the body, the parts of energy moves along
curled paths, and as they change directions, they exert forces on
their surroundings. The force from each part is given by the time
derivative of its momentum:

$$\frac{d\mathbf{p}_n}{dt} = \frac{d\mathbf{p}_n}{ds}\frac{ds}{dt},$$

where ds is a length element in the direction of $\mathbf{p}_n$. Thus the first
term on the right is the directional derivative of $\mathbf{p}_n$ in the direction
of $\hat{\mathbf{p}}_n$, which can be written as $d\mathbf{p}_n/ds = (\hat{\mathbf{p}}_\mathbf{n}\nabla)\mathbf{p}_n$, and the secon
term is simply the speed of the energy movement, i.e. $ds/dt = c$.
Hence

$$\frac{d\mathbf{p}_n}{dt} = c(\hat{\mathbf{p}}_\mathbf{n}\nabla)\mathbf{p}_n.$$

We can also think of it as the *convective derivative* of a momentum flow with velocity, $\mathbf{v} = c\hat{\mathbf{p}}_n$.

$$\frac{d\mathbf{p}_n}{dt} = \frac{\partial \mathbf{p}_n}{\partial t} + \mathbf{v} \cdot \nabla \mathbf{p}_n.$$

Since the flow is presupposed to be stationary, the partial derivative is zero, and we end up with the same result.

The change in momentum per per unit time represents a force, and with $\mathbf{p}_n = p_n\hat{\mathbf{p}}_\mathbf{n}$, we obtain:

$$\mathbf{F}_n = \frac{c}{p_n}(\mathbf{p}_\mathbf{n}\nabla)\mathbf{p}_n$$

By the mathematical identity:

$$(\mathbf{A}\nabla)\mathbf{A} = \frac{1}{2}\text{grad}\,|\mathbf{A}\cdot\mathbf{A}| - \mathbf{A}\times\text{curl}\,\mathbf{A}, \qquad (10.14)$$

we obtain (see also Eq. 4.17):

$$
\begin{aligned}
\mathbf{F}_n &= \frac{c}{p_n}(\frac{1}{2}\text{grad}\,p_n{}^2 - \mathbf{p}_n\times\text{curl}\,\mathbf{p}_n) \\
&= c\text{grad}\,p_n - c(\hat{\mathbf{p}}_n\times\text{curl}\,\mathbf{p}_n) \\
&= c\text{grad}\,\frac{e_n}{c} - c(\hat{\mathbf{p}}_n\times\text{curl}\,\mathbf{p}_n) \\
&= \frac{e_n}{c^2}(-c\text{grad}\,c) + [\text{grad}\,e_n - c(\hat{\mathbf{p}}_\mathbf{n}\times\text{curl}\,\mathbf{p}_n)].
\end{aligned}
$$

The force acting on the whole body is the sum of all the forces acting on each energy packet. We obtain:

$$\mathbf{F} = -\sum_{n=1}^{m}\left[\frac{e_n}{c^2}(c\text{grad}\,c)\right] + \sum_{n=1}^{m}[\text{grad}\,e_n - c(\hat{\mathbf{p}}_\mathbf{n}\times\text{curl}\,\mathbf{p}_n)].$$

First consider a material body in a space with constant c. Then $\text{grad}\,c$ is zero and the first terms on the right hand side of the equation above vanishes. The condition that the energy is confined and no resulting forces are acting on the body, is satisfied if the second terms add up to zero. Thus it is evident that the second terms represent internal forces that cancel out making the resulting force equal to zero.

Now, let the same body be situated in a space with varying c. The internal forces are not affected and still add up to zero as above, while the left term in the sum is no longer zero. We obtain:

$$\mathbf{F} = -(c\operatorname{grad}c)\sum_{n=1}^{m}\frac{e_n}{c^2}$$
$$= -M(c\operatorname{grad}c).$$

We notice that if we place our model of a material body in a field with varying c, it is acted upon by a force in the direction of decreasing c. If we define a new vector $\mathbf{g}$ given by:

$$\mathbf{g} = -c\operatorname{grad}c,$$

the formula above reduces to

$$\mathbf{F} = M\mathbf{g}. \tag{10.15}$$

The vector $\mathbf{g}$ has the dimension of acceleration and can be interpreted as the *acceleration of gravity*, and (10.15) can be interpreted as *Newton's second law of motion*. If the body is kept at rest it will be acted on by the force, F, and if it is free to move, it will get an acceleration, $\mathbf{g}$, that is independent of its mass.

Even if I have simplified the test body to only contain a stationary flow of energy, which clearly is not the case within a real body, I can see no reason why the conclusion should not be valid also for a real body, because the model predicts that absolutely all energy within a material body basically moves as transversal waves with the speed of light.

By (10.13) we saw that another mass, say M_1, creates around itself a $(c\operatorname{grad}c)$-field given by

$$c\operatorname{grad}c = -\frac{GM_1}{r^2}\hat{\mathbf{r}}. \tag{10.16}$$

Let us dub our test mass M_2 and place it at a distance r from M_1. Then by (10.15) M_2 will be acted upon by a force given by:

$$F = G\frac{M_1 M_2}{r^2} \tag{10.17}$$

which we recognize as *Newton's Gravitational Law*. If we on the other hand place M_1 in the field created around M_2, we see that also M_1 is acted upon with the same force in the direction of M_2. Hence F is an attracting force between the two bodies.

Notice, however, that this cannot be the final say about gravity, because I have made some assumptions that may not always be true, mainly stemming from the assumption that the trace of the stress energy tensor gives a uniform stress in all directions. Only a gravitational theory that uses the full power of the stress energy tensor, can be valid in all situations. It can, though, explain some properties that were difficult to see by Newtonian gravity, e.g. the bending of light passing by the sun from far away stars, the redshift of light from heavy stars, why an atomic watch speed up when moved to a higher altitude on earth, etcetera.

10.5   A numerical comparison

The results from this considerations puts us in position to estimate the numerical values of the fundamental constants μ_s, λ_s, and ρ_s. In order to achieve this goal, however, I need to estimate the relation between λ_s and μ_s. I have several times indicated that the speed of longitudinal waves are about the double of transversal waves, and also hinted that the relation even might be exactly so. The assumption that $c_g = 2c_l$ reduces (10.6) to

$$G = \frac{H^2}{192\pi\rho_s}$$

$$\rho_s = \frac{H^2}{192G\pi}.$$

The relation between μ_s and ρ_s is:

$$\mu_s = c^2\rho_s.$$

Newton's gravitational constant, $G = 6.673 \times 10^{-8}$ cm^3/g sec^2. The Hubble Space Telescope Key Project Team has measured Hubble's constant with an uncertainty of 10 percent to $H = 70$ km s^{-1} Mpc$^{-1} = 2.3 \times 10^{-18}$ sec^{-1}. The speed of light $c = 3.0 \times 10^{10}$ cm/sec. With these values inserted the constants amount to

$$\rho_s \approx 1.1 \times 10^{-31} \text{ g/cm}^3,$$
$$\mu_s \approx 1.0 \times 10^{-10} \text{ g/sec}^2 \text{ cm}.$$

The mean density of matter in a flat universe is calculated to about 10^{-30} g/cm^3, which makes it ten times as big as the estimated value of ρ_s. It is remarkable that the two properties seems to be of about the same order of magnitude.

Part V

Relativity

Chapter 11

Principle of Relativity

So God made the dome and separated the waters that were under the dome from the waters that were above the dome. And it was so. (Genesis 1:7)

The negative results on all tests to measure the movement through the spatial continuum can be explained if all lengths in the direction of movement are shortened and our clocks run slower when in motion. However, if it is impossible to measure movements through space, then we have no possibilities to even detect the existence of the spatial continuum itself. So what shall we do? The only viable option is to apply Occam's razor and forget all about absolute motion through space, and only measure relative velocities between observable bodies. This leads directly to Special relativity. Next we register that light from heavy stars is red shifted. Since all periods of the radiation sent out from a star arrive to Earth and can be counted, we have got to deduce that time runs slower at the surface of the star than farther out as also has been detected by direct measurements at different altitudes on Earth.

If we define a time unit as a period of radiation from a certain atom and the length unit as the corresponding wavelength, the *measured speed of light* is one point zero everywhere independent of the *real speed of light*. In a space with varying light speed, VLS, light will be deflected according to Fermat's principle of least time, which corresponds to movements along geodesics. Since material bodies basically are confined wave movements with waves moving in the same way as light, they will also move along geodesics in space just as GR predicts.

11.1   Special Relativity

So far I have assumed that we have access to a measuring system with universal length, time and mass units. If so, with all our sophisticated measuring devices, it should be a simple matter to detect our speed of movement through the spatial continuum. After all, the Earth moves around the Sun, and the Sun moves around the galactic center, so some movements there should be. That was what Albert Michelson and Edward Morley thought when they in 1887 performed their famous experiment, which later has been known as *The Michelson-Morley experiment.* In principle they sent out a beam of light which they split into two beams, one beam straight ahead and one beam normal to it. After having travelled for a certain measured distance, the beams were reflected by two mirrors back to an interferometer where they were made to interfere in a constructive and destructive manner. Simple geometry told them that when the set-up, which was mounted on a millstone flouted in a pool of mercury, was turned around in different directions, the beams of light would have to move different distances when travelling in the direction of movement and transverse to it. That should result in a situation where the interference fringes moved laterally to some extent depending on the direction of the original beam of light in relation to the earth's movement. But, alas, the lines did not move at all, even if the experiment was performed time and time again with increasing accuracy for years to come, no *"aether wind"* could be detected. The negative results are generally considered to be the first strong evidence against the then prevalent aether theory, and initiated a line of research that eventually led to Special relativity, in which the stationary aether concept has no role. See [16] which gives an excellent description of how the experiment was performed.

The negative result of the experiment was soon explained by a contraction of a body, e.g. a meter stick, in the direction of movement. Traditionally this concept was first proposed by George FitzGerald in 1889 and later by Hendrik Lorentz in 1892. I shall not comment on the historical truth of this statement, but surely such a contraction of the meter stick in the direction of movement and a corresponding not changing length of the stick transverse to the direction of movement, can well explain the negative result of the M-M experiment. So farewell to the conception of a universal meter stick.

Then what about time dilatation? There are some direct proves for time dilatation, especially the Hafele and Keating experiment where atomic clocks were flown in opposite directions around the world and compared (see notes on the subject from Georgia State university [8]). The experiments showed clearly that the clocks are slowed down with increasing velocity by very near the amount GR predicts[1].

How then can I propose a model of space and matter that definitely implies a more or less fixed spatial continuum wherein all matter recedes? In reality it all boils down to the invariance of Maxwell's equations under Lorenz transformations. We all live in a world where matter is built up of particles kept together by electromagnetic forces which are proven to be invariant under Lorenz transformations. Hence our world is fully Lorentzian and so are our meter sticks and watches. They change according to movement in space and forbid us to do universal measurements. We have no access to the set of universal length and time units needed do these kind of measurements.

So, is there really no way to detect our speed through space? Well, the COBE anisotropy test gives us a clue (see page 13), but this is only an indirect measurement, by many interpreted as another support for the principle of relativity – it only shows our

---

[1]The length contraction is: $l' = l \cdot \sqrt{1 - v^2/c^2}$, and the time dilatation is: $t' = t/\sqrt{1 - v^2/c^2}$.

relative motion to the bulk of bodies in the universe. There should all the same be a possible way out of this conundrum. Electromagnetic fields are of solenoidal origin while gravitational fields seems to be of irrotational origin. The last ones are probably not Lorenz invariant, so if we could measure gravitational waves with sufficient accuracy, it should perhaps be possible to construct an apparatus that would reveal our speed through the spatial continuum.

However, for the time being we can do quite well with *Special Relativity*, SR, which states that movements in space can only be measured as relative movements between bodies, with all its implications. So far, no measurements that we can presently do, can come in conflict with what we observe because such a discrepancy would provide us with the means to detect an absolute speed through the spatial continuum. (See also page 34 and 70.)

11.2   General Relativity

In Part IV we saw that in this model gravitation implies a space with varying density and wave speeds with decreasing wave speed as we approach a gravitating body. How our meter sticks and clocks behave in such a space is not easy to tell, but let us give it a try.

It is well known that light emitted from a heavy star is red-shifted. As no cycles of oscillation can be lost under way, any atomic clock based on spectral lines emitted from a reliable source will run slower when the radiation is red-shifted. This effect has even been demonstrated on Earth by comparison between extremely precise clocks placed at the ground and the top level of a 22.6 meter high tower. The experiment was performed in the early 1960's by Rebka, and Snyder at the Jefferson Physical Laboratory at Harvard [7]. When in the lower position were the gravitational field is stronger, the clock runs slower than in the upper position. This effect tells us one important property of gravitation. A clock on a heavier celestial body than the Earth runs slower than a corresponding clock on Earth. To see that we can compare spectral lines emitted from known elements on a heavy star with corresponding spectral lines on Earth. As no cycles of oscillation can be lost under way, any atomic clock based on such spectral lines will run slower when the light is red-shifted.

So let us define our standard time unit as one period of light emitted from a reliable atom under a given condition, e.g. caesium-133. Next we define our standard length unit as one wavelength of the same radiation. Then the *measured light speed* becomes one length unit per time unit, i.e. $c = 1.0$, everywhere, even if the *real light speed* should happen to vary throughout space.

Now let us try to compare measurements on a light moon circling a very heavy planet. A woman on the moon communicates with her husband down at the planet. They will readily agree that Bill's clock on the planet runs slower than Eve's clock on the moon, but what about their meter sticks? It turns out that the answer to that question depends upon the relation between the slowdown of the real light speed and the time dilatation on the planet compared to on the moon. The standard length unit may be longer, equal or shorter on the planet than on the moon (see 11.1).

Now let Eve enter a spaceship and travel around in space. She naturally wants to know where she is at any time. But alas, she neither have a working log nor any fixed way-points to fix her positions in relation to. In addition she has got to realize that her meter stick changes under way, so if she try to apply ordinary Euclidean geometry, she is in great trouble. Even if she could find way-points. The measured distance between them could be different depending on the path of measurement, and besides she couldn't be on both points at the same time. For example, a diameter calculated based on the orbit around the planet would be different from a distance measured along the diameter. If the journey had taken place on a plane, she could have concluded that the plane was not flat, but had a dip in it around the star so the distances along that path had to be longer. It is much harder to see that a three dimensional space can be curved in a similar way, but now she has the choice of working with a variable standard meter, or defining the space itself as being curved. According to GR the last option is the preferred one, but then she would need a new kind of geometry to do her

Let the length unit be defined as a wavelength of light emitted from ceasium-133, and the time unit as a period. Then the *measured light speed* is given by

$$c = \lambda/T = 1.0, \tag{11.1}$$

independent of the *real light speed*.

The length unit on the light moon and the heavy planet are λ_m and λ_p respectively.

Light from heavy stars are red-shifted, hence a period on the planet is longer than on the moom, $T_p > T_m$:

$$T_p = \alpha T_m, \quad \text{where} \quad \alpha > 1.$$

In a VSL-theory of gravitation the real light speed near a heavy planet is lower than on a light moon, so: $c_p < c_m$.

$$c_p = \beta c_m, \quad \text{where} \quad \beta < 1.$$

A wavelength on the surface of the planet is given by:

$$\begin{aligned} \lambda_p \quad &= \alpha\beta \, c_m T_m \\ &= \alpha\beta \, \lambda_m. \end{aligned}$$

It is longer, equal, or shorter than the wavelength on the moon depending on the values of α and β. Hence our meter sticks may or may not vary throughout space.

Figure 11.1: Length measurements in space.

calculations in.

We now have several variables to account for. The standard meter stick will vary in the x, y, and z directions depending on our velocity and direction of movement through space, and besides, our time unit will also vary with velocity. The only reference points to determine our position in relation to are other celestial objects, which themselves are in motion. In addition the real speed of light will vary with the density of the spatial continuum. To cope with all that, we would need a set of numbers at every place we visit in space to tell us our position in relation to other objects at a specific time, the spatial condition at that point, and how we move. At the time of Einstein a geometry that could manage all that was already proposed by Bernhard Riemann (1826–1866) in his inaugurational lecture (1854) in Göttingen 'Ueber die Hypothesen, welche der Geometrie zu Grunde liegen' (On the Hypotheses which lie at the Bases of Geometry), published 2 years after his death in 1868 [13]. The mathematics behind this elaborate geometry was developed further by the Italian mathematician Gregorio Ricci-Curbastro (1853 – 1925) and has been called Ricci calculus after him [19]. In such a geometry the condition at any point in space and time can be described by the 16 numbers – whereof 10 different – in a (0,2) symmetric tensor, the so called *metric tensor*, $g_{\mu\nu}$.

Here it will be too far-fetched to try to go into the development of the necessary geometry to describe the modern geometry of space-time in the so called *metric space*. However, while Riemannian metrics had only positive terms, what was now needed was a geometry that also included negative terms. This geometry has been called pseudo-Riemannian geometry. Here the metric has both positive and negative terms, and in GR the so called signature of the metric is usually chosen with time (t) being negative and the spatial parts $(x,\,y,\,z)$ being positive, i.e. $(-+++)$ signature.

We cannot see the spatial continuum and have no way to measure how we move through it, but that does not mean that it is non-existent. It seems like we are doomed to live in a relativistic world, so in the two last chapters of this book I have tried to figure out how the mechanical model of space and matter can be given a form that fits into the theory of relativity. I do not know to what extent I have succeeded, but I think I have found some interesting connections.

Barcel and G. Jannes at Instituto de Astrofisica de Andalucia have written an interesting paper where they show that even an internal observer in a fish bowl would experience a relativistic reality if she had to do her observations entirely by acoustical phenomena.

They wrote [2, 3]:

> As a matter of fact, in our Gedankenmodel, we showed
> that this fundamental external world can even be New-
> tonian (the laboratory), while still reproducing a rela-
> tivistic behaviour with respect to an internal observer.
> Only if this internal observer were able to probe the
> external world, e.g. through observation of phenomena
> linked directly to a quantum regime of gravity, would he
> be able to really know anything about the fundamental
> physics.

Chapter 12

Special relativity

Special relativity is taken care of by the Lorenz transforms. We are all passenger in a Lorentzian spacecraft, where both we and our measuring equipments and units are subject to Lorentz transformations, and in Chap. 4 I found that in such a frame, the electromagnetic equations are invariant. We cannot register that we probably are travelling through space at an immense speed compared to what we are travelling around with inside our own spacecraft – for example in our solar system. So far I have assumed that we have access to a measuring system with universal length, time and mass units. In the real world, however, such measuring devices are not obtainable. Our clocks may vary depending on the speed with which we are travelling through space and even our measuring sticks may vary depending on the direction of movement. This chapter is meant as a sketch to see how the principle of relativity might be implemented into this model of space and matter.

12.1 The Michelson/Morley experiment.

In the nineteenth century it was commonly believed that light was waves in a lumeniferous aether. If so, it should be possible to set up an apparatus to measure the speed of the earth through the aether. Thus Albert A. Michelson and Edward E. Morley set up an

experiment that was precise enough to detect the motion. It was
an interferometer consisting of two rods normal to each other which
could be turned so the rods were shifted between being pointing
along the movement and transverse to it. A beam of light was split
into two beams along each of the rods and reflected by mirrors at
the end of the rods and back to the interferometer. The idea was
that the two beams would have to travel different distances along or
across to the movements, and that the fringes in the interferometer
would shift positions as the apparatus was turned around, but no
shifts was observed (see Fig. 12.1). The negative result of the
Michelson/Morley experiment was the first nail in the coffin that
eventually buried the old aether theory.

The proposed model, however, is a successor of the same ideas,
so the negative result has got to be explained in some way. George
FitzGerald and Henrik Antoon Lorenz found that the negative result could be explained if the lengths of the rod in the direction of
movement was contracted to some extent, but that was considered
to be an ad hoc hypothesis.

The length contraction can be derived from the Lorentz transformation:

$$x' = \frac{x - vt}{\sqrt{1 - v^2/c^2}},$$

$$t' = \frac{t - vx/c^2}{\sqrt{1 - v^2/c^2}}.$$

If the length in the resting frame is L, then the length in the moving
frame as observed by an observer in the resting frame is given by[1]:

$$L' = L\sqrt{1 - v^2/c^2}.$$

The observer, however, is restricted to observe the contraction with
light, which moves with the speed c. Hence she cannot see the tip
and the tail of the measured stick at the same time, but rather as
two events at different space-times.

---

[1]The result has been derived time and time again along different paths since
Lorenz first did it in the fall of the nineteenth century, so I shan't bother to take
it up here. The result has always been the same.

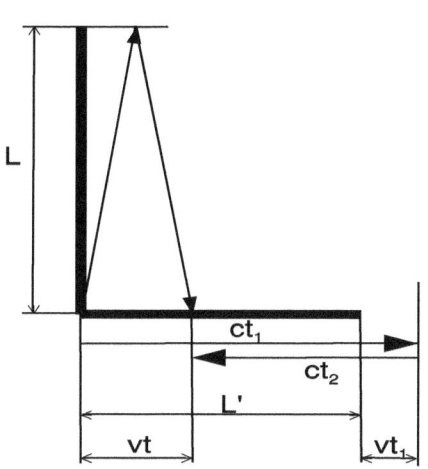

Figure 12.1:
Michelson/Morley experiment. *The interferometer is moving to the right with a velocity v. Michelson-Morley found that there was no change in the interference as the apparatus was turned 90°. The only possible explanation if the proposed model is true, is that the rod in the direction of movement is shortened to some extent in relation to the transverse rod.*

Let in stead a *daemon* with access to universal measuring devices, who can observe the positions of all the points on an object at the same time, do the observation. The result would be the real contraction and not just the observed one. He sets up an interferometer with two equally long arms, L, as shown in Fig. 12.1. First he test the apparatus while it is at rest. He finds that the light beams move back and forth along the arms in a time, t:

$$ct = 2L,$$
$$t = 2L/c.$$

Then he sets the apparatus into a movement along one of the arms with a speed, v. The beam would travel the length of the transverse beam and back again in a time, t', given by:

$$(2L)^2 + (vt')^2 = (ct')^2,$$
$$t' = \frac{2L}{c\sqrt{1 - v^2/c^2}}. \tag{12.1}$$

The beam in the direction of movement will travel a length that perhaps has changed to L', until it reaches the mirror in time, t'_1, and back again to the interferometer in a time t_2'. For the interferometer to show no difference between the two beams, the sum of these two times has got to be like t'. So the daemon sets up the three equations:

$$\begin{aligned}
t' &= t_1' + t_2', \\
ct_1' - ct_2' &= vt', \\
L' + vt_1' &= ct_1'.
\end{aligned}$$

He solves the first equation with respect on t_2' and plug it into the next equation, which he solves with respect on t_1'. Finally he plugs t_1' into the last equation and obtain the resulting time, t':

$$t' = \frac{2L'}{c(1 - v/c)}.$$

He now has two expression for the same time and can find if L has changed:

$$\frac{2L}{c\sqrt{1 - v^2/c^2}} = \frac{2L'}{c(1 - v/c)},$$

$$L' = L\,\frac{1 - v/c}{\sqrt{1 - v^2/c^2}}.$$

If $v/c \ll 1$, this expression can be reformulated a bit and we obtain:

$$\begin{aligned}
L' &\approx L\,(1 - v/c)\left(1 + \frac{1}{2}v^2/c^2\right) \\
&\approx L(1 - v/c).
\end{aligned}$$

Clearly $L' < L$, so we have a real contraction in the direction of motion and that even more than SR predicts. We also notice that $t' > t$ while we would not have seen any difference in our own spaceship observations. We are therefore forced to accept that our clocks run slower when we are in motion.

These two examples illustrate the difference between relativistic and mechanical approaches. The relativistic approach is the only one that gives us testable results. We are not daemons and have no access to universal test equipments, and moreover, we are dependent of light to observe what is going on around us, so the result our daemon found cannot be tested and is therefore completely useless. Therefore we have no options but to accept the relativistic approach with all its difficult implications.

12.2 Lorenz invariance and relativity

In Chapter 4 I found that all solenoidal elastodynamic equation and hence electrodynamic equations can be expressed in a form that are invariant under Lorenz transformations. It is also well documented that binding forces between atoms are of electromagnetic origin. It is therefore natural to conclude that material properties will be such that their characteristics are invariant under Lorentz transforms. Our meter sticks will shorten in the direction of movements, and our clocks will slow down when in motion. This makes it impossible to detect which Lorenz frame we do our observations in, because they are all like, even if the frame does not move at all. In that special case our measuring devices would be universal and like our demon's. It is with basis in that frame I have developed the four-tensors and relations that are frame independent, which we can trust will be the same in what present frame we should happen to live in.

Gravitational forces, however, are of irrotational elastodynamic origin, and therefore not invariant under Lorenz transformations. If it should become possible to measure the speed of gravitational waves (provided that gravitational waves proves to be longitudinal pressure waves) with sufficient precision, it should open for a test of the mechanical model contra the principle of relativity.

Chapter 13

General relativity

So far I have assumed that space is homogeneous and the light speed is the same anywhere. Our clocks may vary depending on the speed with which we are travelling through space and even our measuring sticks may vary depending on the direction of movement, but they have not varied from place to place in space. In the chapters about Newtonian gravity I showed that gravity could be explained as an effect of varying light speed[1]. Here I'll try a slightly different approach. John Wheeler said: "Matter tells space how to curve, and space tells matter how to move." Matter is best described by the solenoidal stress energy tensor (Sec. 3.10). To describe space fully we would need a corresponding irrotational stress energy tensor (Sec. 3.11). The interaction between matter and space should be governed by Newton's third law of motion: "For every action, there is an equal and opposite reaction." Hence the sum of these two tensors should be zero. To find such a relation in an expanding space, will be the topic of this chapter.

---

[1]Variable speed of light (VSL) theories has been proposed by several scientists, notably by Einstein himself in 1911, who certainly dismissed it later when he formulated his principle of general relativity.

13.1   D'Alembert's operator

Recall the Navier-Cauchy equation (see Sec. 3.4):

$$c_g{}^2 \text{grad div } \mathbf{u} - c_l{}^2 \text{curl curl } \mathbf{u} + \frac{\mathbf{b}}{\rho_s} = \ddot{\mathbf{u}}. \qquad (13.1)$$

The displacement field can be decomposed into two properties:

$$\mathbf{u} = \hat{\mathbf{u}} + \tilde{\mathbf{u}},$$

where

$$\begin{aligned}
\hat{\mathbf{u}} &= -\text{grad } \phi, \\
\tilde{\mathbf{u}} &= \text{curl } \psi, \qquad \text{div } \psi = 0.
\end{aligned}$$

There are no external forces acting in the spatial continuum, so the body force, $\mathbf{b}$, is set to zero.
By the identity:

$$\nabla^2 \mathbf{A} = \text{grad div } \mathbf{A} - \text{curl curl } \mathbf{A},$$

and that both curl $\hat{\mathbf{u}}$ and div $\tilde{\mathbf{u}}$ vanish, the Navier-Cauchy equation can be written:

$$\begin{aligned}
c_g{}^2 \nabla^2 \hat{\mathbf{u}} + c_l{}^2 \nabla^2 \tilde{\mathbf{u}} + &= \frac{\partial^2 \hat{\mathbf{u}}}{\partial t^2} + \frac{\partial^2 \tilde{\mathbf{u}}}{\partial t^2}, \\
c_g{}^2 \nabla^2 \hat{\mathbf{u}} - \frac{\partial^2 \hat{\mathbf{u}}}{\partial t^2} &= -\left(c_l{}^2 \nabla^2 \tilde{\mathbf{u}} - \frac{\partial^2 \tilde{\mathbf{u}}}{\partial t^2} \right). \qquad (13.2)
\end{aligned}$$

In fact we have got to deal with two slightly different wave operators, which we can depict:

$$\begin{aligned}
\hat{\Box} &= \hat{\partial}_\nu \hat{\partial}^\nu = c_g{}^2 \nabla^2 - \frac{\partial^2}{\partial t^2}, \\
\tilde{\Box} &= \tilde{\partial}_\mu \tilde{\partial}^\mu = c_l{}^2 \nabla^2 - \frac{\partial^2}{\partial t^2}.
\end{aligned}$$

The last one we recognize as d'Alembert's operator. They describe the displacement amplitudes of two waves:

$$c_g{}^2 \nabla^2 \, \hat{\mathbf{u}} - \frac{\partial^2 \hat{\mathbf{u}}}{\partial t^2} = 0,$$

$$c_l{}^2 \nabla^2 \, \tilde{\mathbf{u}} - \frac{\partial^2 \tilde{\mathbf{u}}}{\partial t^2} = 0,$$

namely P- and S-waves respectively. The Navier-Cauchy equation can compactly be written:

$$\hat{\Box} \, \hat{\mathbf{u}} = \tilde{\Box} \, \tilde{\mathbf{u}}.$$

From these equations we can derive two stress energy tensors, one for irrotational and one for solenoidal deformations.

13.2   The stress energy tensors

In Chapter 3, Sec. 3.30, I found that the complete stress energy tensor, **T**, may be expressed as the sum of solenoidal and irrotational tensors, We obtained:

$$T_\alpha{}^\beta = \hat{T}_\alpha{}^\beta + \tilde{T}_\alpha{}^\beta .$$

In the absence of any outer forces represented by **b** in the N-C equation, the sum of the two tensors above becomes zero, hence:

$$\hat{T}_\alpha{}^\beta = -\tilde{T}_\alpha{}^\beta .$$

This may be interpreted as Newton's third law: *To every action there is always opposed an equal reaction.* Matter, in the disguise of electromagnetic energy, performs the action, and the reaction are forces caused by irrotational deformations.

As with the Navier-Cauchy equation above, (13.2), we can now apply the two wave operators on the stress energy tensors. We obtain:

$$c_g{}^2 \nabla^2 \, \hat{\mathbf{T}} - \frac{\partial^2 \hat{\mathbf{T}}}{\partial t^2} = -\left(c_l{}^2 \nabla^2 \, \tilde{\mathbf{T}} - \frac{\partial^2 \tilde{\mathbf{T}}}{\partial t^2} \right),$$

$$c_g{}^2 \nabla^2 \, \hat{T}_\alpha{}^\beta + c_l{}^2 \nabla^2 \, \tilde{T}_\alpha{}^\beta = \frac{\partial^2 \, T_\alpha{}^\beta}{\partial t^2} . \tag{13.3}$$

13.3   The solenoidal stress energy tensor

The mixed solenoidal stress energy tensor is given by (see Sec. 3.10):

$$
\tilde{T}^{\nu}_{\mu} = \begin{bmatrix} -\tilde{e} & -\tilde{S}_x/c & -\tilde{S}_y c & -S'_z/c \\ \tilde{S}_x/c & -\tilde{\sigma}_{xx} & -\tilde{\sigma}_{xy} & -\tilde{\sigma}_{xz} \\ \tilde{S}_y/c & -\tilde{\sigma}_{yx} & -\tilde{\sigma}_{yy} & -\tilde{\sigma}_{yz} \\ \tilde{S}_z/c & -\tilde{\sigma}_{zx} & -\tilde{\sigma}_{zy} & -\tilde{\sigma}_{zz} \end{bmatrix},
\tag{13.4}
$$

with, $\tilde{\sigma}_{ab} = -\delta_{ab}\tilde{e} + \rho_s \dot{\tilde{u}}_a \dot{\tilde{u}}_b + \mu_s(\operatorname{curl}\tilde{u})_a(\operatorname{curl}\tilde{u})_b$. Here $\tilde{e}$ is the energy density, $\tilde{S}$ the energy flow vector, $\tilde{\sigma}$ Maxwell's stress tensor, and c_l the speed of light.

The trace of the tensor becomes:

$$
\begin{aligned}
\operatorname{tr}\hat{T} &= -\tilde{e} - \sum_{a=1}^{3}[-\delta_a^a \tilde{e} + \rho_s \dot{\tilde{u}}_a \dot{\tilde{u}}_a + \mu_s(\operatorname{curl}\tilde{u})_a(\operatorname{curl}\tilde{u})_a] \\
&= -\tilde{e} + 3\tilde{e} - 2\tilde{e}_k - 2\tilde{e}_p = 0
\end{aligned}
$$

The kinetic and potential parts of the energy density are $\tilde{e} = \tilde{e}_k + \tilde{e}_p$ respectively.

If $\tilde{\sigma}_{xx} = \tilde{\sigma}_{yy} = \tilde{\sigma}_{zz}$, the pressure $\tilde{p}$ becomes:

$$
-\tilde{p} = \tilde{\sigma}_{xx} = \tilde{\sigma}_{yy} = \tilde{\sigma}_{zz}.
$$

Since the trace vanishes, we obtain:

$$
\tilde{p} = \frac{1}{3}\tilde{e}.
$$

In an expanding space, a surface around a sphere with radius, r, will recede from the origin with a speed, $v = H \cdot r$. The confined energy within the sphere will loose energy at a pace given by the force acting on the surface, $a = 4\pi r^2$, times the radial velocity, v. The pressure on the surface is given by $p = 1/3\,e$, and the energy lost per unit time becomes:

$$
\begin{aligned}
\frac{\partial E}{\partial t} &= -a \cdot p \cdot v \\
&= -4\pi r^2 \cdot \frac{1}{3}e \cdot Hr, \\
\frac{1}{4/3\,\pi r^3}\frac{\partial E}{\partial t} &= -He,
\end{aligned}
$$

so

$$\frac{\partial e}{\partial t} = -He. \tag{13.5}$$

The second derivative with respect on time is:

$$
\begin{aligned}
\frac{\partial^2 e}{\partial t^2} &= -H\frac{\partial e}{\partial t} \\
&= -H(-He) \\
&= H^2 e.
\end{aligned}
$$

Here I have assumed that H is constant, but it wouldn't make any fundamental difference if H is a function of time. For example if $H = \text{const.}/t$, then: $\partial^2 e/\partial t^2 = 2H^2 e$.

All terms in a stress energy tensor are dependent on the energy and are reduced by the same amount. Hence we can extend the formula above to:

$$\frac{\partial^2}{\partial t^2}T_\alpha{}^\beta = H^2 T_\alpha{}^\beta. \tag{13.6}$$

13.4 The irrotational stress energy tensor

The mixed irrotational stress energy tensor is given by (3.26):

$$
\hat{T}^\nu_\mu =
\begin{bmatrix}
-\hat{e} & -\frac{\hat{S}_x}{c_g} & -\frac{\hat{S}_y}{c_g} & -\frac{\hat{S}_z}{c_g} \\
\frac{\hat{S}_x}{c_g} & \hat{\sigma}_{xx} & \hat{\sigma}_{xy} & \hat{\sigma}_{xz} \\
\frac{\hat{S}_y}{c_g} & \hat{\sigma}_{yx} & \hat{\sigma}_{yy} & \hat{\sigma}_{yz} \\
\frac{\hat{S}_z}{c_g} & \hat{\sigma}_{zx} & \hat{\sigma}_{zy} & \hat{\sigma}_{zz}
\end{bmatrix},
\tag{13.7}
$$

with, $\hat{\sigma}_{ab} = \rho_s(\dot{\hat{u}}_a\dot{\hat{u}}_b) + \delta_{ab}(\hat{e}_p - \hat{e}_k)$. The kinetic and potential energy of the irrotational field are set to $\hat{e}_k = 1/2\rho_s\dot{\hat{u}}^2$, and $\hat{e}_p = 1/2(\lambda_s + 2\mu_s)(\text{div } \hat{u})^2$, respectively.

First I write the diagonal components of the mixed tensor (13.7):

$$\begin{aligned}
\hat{T}^0_0 &= -\hat{e}, \\
\hat{T}^1_1 &= \rho_s \dot{\hat{u}}_x \dot{\hat{u}}_x - \hat{e}_k + \hat{e}_p, \\
\hat{T}^2_2 &= \rho_s \dot{\hat{u}}_y \dot{\hat{u}}_y - \hat{e}_k + \hat{e}_p, \\
\hat{T}^3_3 &= \rho_s \dot{\hat{u}}_z \dot{\hat{u}}_z - \hat{e}_k + \hat{e}_p,
\end{aligned}$$

and then split it up into two more terms. We obtain

$$\begin{aligned}
\hat{T}^0_0 &= -(\hat{e}_p + \hat{e}_k) - (\hat{e}_p - \hat{e}_k) + (\hat{e}_p - \hat{e}_k), \\
\hat{T}^1_1 &= \rho_s \dot{\hat{u}}_x{}^2 + (\hat{e}_p - \hat{e}_k), \\
\hat{T}^2_2 &= \rho_s \dot{\hat{u}}_y{}^2 + (\hat{e}_p - \hat{e}_k), \\
\hat{T}^3_3 &= \rho_s \dot{\hat{u}}_z{}^2 + (\hat{e}_p - \hat{e}_k),
\end{aligned}$$

Now we can define two new temporary tensors,

$$R'^{\,\nu}_{\mu} = \begin{bmatrix}
-2\hat{e}_p & -\hat{T}^1_0 & -\hat{T}^2_0 & -\hat{T}^2_0 \\
\hat{T}^0_1 & \rho_s \dot{\hat{u}}_x{}^2 & \hat{T}^2_1 & \hat{T}^3_1 \\
\hat{T}^0_2 & \hat{T}^1_2 & \rho_s \dot{\hat{u}}_y{}^2 & \hat{T}^3_2 \\
\hat{T}^0_3 & \hat{T}^1_3 & \hat{T}^2_3 & \rho_s \dot{\hat{u}}_z{}^2
\end{bmatrix},$$

and

$$(\hat{e}_p - \hat{e}_k)\eta_{\mu}{}^{\nu} = \begin{bmatrix}
(\hat{e}_p - \hat{e}_k) & 0 & 0 & 0 \\
0 & (\hat{e}_p - \hat{e}_k) & 0 & 0 \\
0 & 0 & (\hat{e}_p - \hat{e}_k) & 0 \\
0 & 0 & 0 & (\hat{e}_p - \hat{e}_k)
\end{bmatrix}.$$

Hence the stress energy tensor can be written:

$$\hat{T}^{\nu}_{\mu} = R'^{\,\nu}_{\mu} + (\hat{e}_p - \hat{e}_k)\eta_{\mu}{}^{\nu} \tag{13.8}$$

The trace of **R'** is:

$$\begin{aligned}
\operatorname{tr} \mathbf{R'} &= R'^{\,\mu}_{\mu} = -2\hat{e}_p + \rho_s(\dot{\hat{u}}_x{}^2 + \dot{\hat{u}}_y{}^2 + \dot{\hat{u}}_z{}^2) \\
R' &= -2(\hat{e}_p - \hat{e}_k).
\end{aligned}$$

Notice that the trace of $\mathbf{R}'$ is depicted R' without indices. We obtain:

$$(\hat{e}_p - \hat{e}_k)\eta_\mu{}^\nu = -\frac{1}{2}R'\eta_\mu{}^\nu.$$

We plug this into (13.9) and obtain:

$$\hat{T}_\alpha^\beta = R'_\alpha{}^\beta - \frac{1}{2}R'\eta_\alpha{}^\beta. \qquad (13.9)$$

In a free wave the energy is shifting between potential and kinetic energy, so in a free wave the two energy forms are equal, and the difference, $\hat{e}_p - \hat{e}_k$, is zero and so is the trace, i.e. $R' = 0$.

In this model, however, matter is characterized by confined energy that displaces some spatial mass. In that extreme case the left side of the equation above is static, and the kinetic energy is zero. Thus (13.9) describes both a dynamic and a static state, or a combination of these two states.

13.5 Einstein's equation

Consider a volume with confined wave energy. Then by (13.3) we can set up the equation:

$$c_g{}^2\nabla^2\,\hat{T}_\alpha{}^\beta + c_l{}^2\nabla^2\,\tilde{T}_\alpha{}^\beta = \frac{\partial^2\,T_\alpha{}^\beta}{\partial t^2}.$$

In expanding space it reduces to:

$$c_g{}^2\nabla^2\,\hat{T}_\alpha{}^\beta + c_l{}^2\nabla^2\,\tilde{T}_\alpha{}^\beta = H^2 T_\alpha{}^\beta.$$

Now, we only consider the space outside the area with confined energy. Then there will be no solenoidal deformations, and the equation reduces to:

$$c_g{}^2\nabla^2\,\hat{T}_\alpha{}^\beta = H^2 T_\alpha{}^\beta.$$

Finally we plug in (13.9) and obtain:

$$c_g{}^2\nabla^2\left(R'_\alpha{}^\beta - \frac{1}{2}R'\eta_\alpha{}^\beta\right) = H^2 T_\alpha{}^\beta.$$

In Chapter 10 I found the relation between Hubbles constant and the gravitational constant given by (10.6):

$$G = \frac{H^2}{24\pi\rho_s} \frac{c_l^4}{c_g^4},$$

$$H^2 = \frac{3\rho_s c_g^4}{c_l^2} \cdot \frac{8\pi G}{c_l^2},$$

which I plug it into the equation above and order:

$$\frac{c_l^2}{3\rho_s c_g^2} \nabla^2 \left(R'^{\beta}_{\alpha} - \frac{1}{2} R' \eta_\alpha^{\beta} \right) = \frac{8\pi G}{c_l^2} T_\alpha^{\beta}.$$

Now, I define a new tensor from the temporary tensor $\mathbf{R'}$ and obtain:

$$\mathbf{R} = \frac{c_l^2}{3\rho_s c_g^2} \nabla^2 \mathbf{R'},$$

or:

$$R_\alpha^{\beta} = \frac{\mu_s}{3\rho_s(\lambda_s + 2\mu_s)} \nabla^2 \begin{bmatrix} -2\hat{e}_p & -\hat{T}_0^1 & -\hat{T}_0^2 & -\hat{T}_0^2 \\ \hat{T}_1^0 & \rho_s\hat{u}_x^2 & \hat{T}_1^2 & \hat{T}_1^3 \\ \hat{T}_2^0 & \hat{T}_2^1 & \rho_s\hat{u}_y^2 & \hat{T}_2^3 \\ \hat{T}_3^0 & \hat{T}_3^1 & \hat{T}_3^2 & \rho_s\hat{u}_z^2 \end{bmatrix} \quad (13.10)$$

and we obtain:

$$R_\alpha^{\beta} - \frac{1}{2} R \eta_\alpha^{\beta} = \frac{8\pi G}{c_l^2} T_\alpha^{\beta}.$$

This far the equations are developed without needing to take into account whatever metric there might be, but surely there is a yet unknown metric that may be like the Minkowski metric or something else. I set this still unknown metric to $\mathbf{g}$. Its covariant and contravariant components are g_{ij} and g^{kl} respectively. They are inverse to each other such that $g_{ij}g^{jk} = g_i^{k} = \delta_i^{k}$, and they can be used to raise or lover the indices of any tensor. Now I lower the upper indices in the equation above and get:

$$g_{\beta\nu} R_\mu^{\beta} - g_{\beta\nu} \frac{1}{2} R g_\mu^{\beta} = \frac{8\pi G}{c_l^2} g_{\beta\nu} T_\mu^{\beta},$$

$$R_{\mu\nu} - \frac{1}{2} R g_{\mu\nu} = \frac{8\pi G}{c_l^2} T_{\mu\nu}. \quad (13.11)$$

We see that the equation above is similar to the Einstein equation without the cosmological constant. In a region of space with no matter we obtain:

$$R_{\mu\nu} = \frac{1}{2} R g_{\mu\nu},$$

and with R just a number which locally can be replaced by a constant, k, given by $R = 2k$, we obtain the *Einstein condition*:

$$\mathbf{R} = k\,\mathbf{g}.$$

This is an *Einstein manifold* where the Ricci tensor is proportional to the metric. If we should happen to be able to solve the rather intricate set of equations above, we would obtain the metric, $g_{\alpha\beta}$, as well as the Ricci curvature tensor, $R_{\alpha\beta}$. Thus (13.11) becomes precisely like the *Einstein equation*, and the *Einstein tensor* itself becomes:

$$G_{\mu\nu} = R_{\mu\nu} - \frac{1}{2} R g_{\mu\nu} \qquad (13.12)$$

The defined space needs not at all be flat, so from here on we have got to do all our calculations in a curved space time. The electromagnetic equations were also first defined as mixed tensors in Minkowski space time and hence need to be redefined in the same metric space in order to remain valid.

13.6   The metric space

A beam of light will follow a geodesic in metric space, so let us see where that leads us. Before I proceed, I need to take a brief look at some mathematical identities in pseudo Riemannian metric space.

The *covariant derivative* of the vector, $\mathbf{V}$, can be denoted in several ways. I shall alternate between the two notations below:

$$\begin{aligned}
\nabla_m V^n &= \partial_m V^n + \Gamma^n_{rm} V^r, \\
V^n{}_{;m} &= V^n{}_{,m} + \Gamma^n_{rm} V^r.
\end{aligned}$$

Thus a covariant derivative is denoted with a ∇ sign or a semi-colon, while a normal partial derivative is indicated with a ∂ sign or a comma. The connection coefficient, Γ^a_{bc}, is the Christoffel symbol (see below). The covariant derivative of a covariant vector is accordingly:

$$V_{n;m} = V_{n,m} - \Gamma^r_{mn} V_r.$$

The *Christoffel symbol*, Γ^a_{bc}, is a connection coefficient given by[2]:

$$\Gamma^i_{k\ell} = \frac{1}{2} g^{im} \left(\frac{\partial g_{mk}}{\partial x^\ell} + \frac{\partial g_{m\ell}}{\partial x^k} - \frac{\partial g_{k\ell}}{\partial x^m} \right)$$

$$= \frac{1}{2} g^{im} (g_{mk,\ell} + g_{m\ell,k} - g_{k\ell,m}).$$

It is symmetric in the two lower indices, $\Gamma^a_{bc} = \Gamma^a_{cb}$.

Actually there are two variants of the Christoffel symbol. The Christoffel symbol of the second kind above, and the Christoffel symbol of the first kind is given simply by raising one index with the metric in front:

$$\Gamma_{cab} = \frac{1}{2} \left(\frac{\partial g_{ca}}{\partial x^b} + \frac{\partial g_{cb}}{\partial x^a} - \frac{\partial g_{ab}}{\partial x^c} \right)$$

$$= \frac{1}{2} (g_{ca,b} + g_{cb,a} - g_{ab,c}). \tag{13.13}$$

The covariant derivative of the metric tensor, g_{ab} is:

$$g_{ab;c} = g_{ab,c} - \Gamma^r_{bc} g_{ar} - \Gamma^r_{ca} g_{br}$$

$$= g_{ab,c} - \frac{1}{2} g^{rm} (g_{mb,c} + g_{mc,b} - g_{bc,m}) g_{ar}$$

$$\quad - \frac{1}{2} g^{rm} (g_{ma,c} + g_{mc,a} - g_{ac,m}) g_{br}$$

$$= g_{ab,c} - \frac{1}{2} (g_{mb,c} + g_{mc,b} - g_{bc,m}) \eta^m{}_a$$

$$\quad - \frac{1}{2} (g_{ma,c} + g_{mc,a} - g_{ac,m}) \eta^m{}_b$$

$$= g_{ab,c} - \frac{1}{2} (g_{ab,c} + g_{ac,b} - g_{bc,a} + g_{ba,c} + g_{bc,a} - g_{ac,b}),$$

$$g_{ab;c} = 0 \tag{13.14}$$

---

[2]Here m is a dummy index, which is summed over.

The covariant derivative of the metric tensor vanishes.

A *geodesic* is the path a material body or a photon will follow in space-time if not affected by any external force. Consider a vector, **V**, in a metric space, and find how the vector varies along a given curve. We take the directional derivative along the curve and apply the chain rule:

$$\frac{\mathrm{d}V^n}{\mathrm{d}s} = \frac{\mathrm{d}V^n}{\mathrm{d}y^m}\frac{\mathrm{d}y^m}{\mathrm{d}s}$$

$$= \left(\frac{\partial V^n}{\partial y^m} + \Gamma^n_{mr}V^r\right)\frac{\mathrm{d}y^m}{\mathrm{d}s}$$

$$= \frac{\partial V^n}{\partial y^m}\frac{\mathrm{d}y^m}{\mathrm{d}s} + \Gamma^n_{mr}V^r\frac{\mathrm{d}y^m}{\mathrm{d}s}$$

$$= \frac{\partial V^n}{\partial s} + \Gamma^n_{mr}V^r\frac{\mathrm{d}y^m}{\mathrm{d}s}$$

The property, $\mathrm{d}y^m/\mathrm{d}s$, has length 1, and it is directed along the curve, so it is a *tangent vector* of unit length to the curve.

Now, let the vector, V^n, happen to be a tangent vector to the curve, then $V^n = a \cdot \mathrm{d}y^n/\mathrm{d}s$, and the above expression takes the form

$$\frac{\mathrm{d}V^n}{\mathrm{d}s} = \frac{\mathrm{d}}{\mathrm{d}s}\left(a\frac{\mathrm{d}y^n}{\mathrm{d}s}\right) + a\Gamma^n_{mr}\frac{\mathrm{d}y^n}{\mathrm{d}s}\frac{\mathrm{d}y^m}{\mathrm{d}s}.$$

If the curve is such that the variation of the tangent vector is zero all the way along the curve, then the curve:

$$\frac{\mathrm{d}^2 y^n}{\mathrm{d}s^2} + \Gamma^n_{mr}\frac{\mathrm{d}y^n}{\mathrm{d}s}\frac{\mathrm{d}y^m}{\mathrm{d}s} = 0, \qquad (13.15)$$

or alternatively

$$\frac{\mathrm{d}^2 y^n}{\mathrm{d}\tau^2} + \Gamma^n_{mr}\frac{\mathrm{d}y^n}{\mathrm{d}\tau}\frac{\mathrm{d}y^m}{\mathrm{d}\tau} = 0, \qquad (13.16)$$

is a *geodesic* in the metric space. A body that is not influenced by any external forces, is supposed to follow a geodesic.

13.7   Fermat's principle of least time

Fermat's principle states that in an optical medium with varying light speed, the path of light taken between two points is the path that can be traversed in the least time. The principle is equivalent with finding the shortest optical length between two points by applying the variational principle:

$$\delta S \;=\; \delta \int_a^b v \, dt = 0,$$

where v is the speed of light in the considered medium.

The problem is that in our local reference frame the speed of light is constant. So let us invent a demon, and let him construct a time unit and a meter stick as well as a mass unit far away from any heavenly objects. Since he is a demon, he can bring with him these units to wherever he wants in space, and measure whatever he likes. We can consider his measurements as being universal, so when we speak of universal measurements, this is what it is. Our own measurements, however, have got to be with our locally constructed units and devices.

Now, let us invite the demon to go with us in spaceship Earth, and agree upon measuring the speed of a massless particle, the photon. The demon measure the speed to be c_l, and we find it to be c. In order to compare this two measurements, we have got to find an invariant property, which happens to be the square of the velocity vector. For us, our space-craft is a metric space, so we need to know the local metric. Hence we obtain:

$$c_l{}^2 \;=\; g_{\mu\nu}\dot{x}^\mu \dot{x}^\nu,$$

$$c_l \;=\; \sqrt{g_{\mu\nu}\dot{x}^\mu \dot{x}^\nu},$$

and Fermat's principle takes the form

$$\delta \int_a^b \sqrt{g_{\mu\nu}\dot{x}^\mu \dot{x}^\nu}\, dt = 0. \tag{13.17}$$

To find the extremal value of this equation we need to apply the *The Euler-Lagrange* differential equation, which is the fundamental equation of *calculus of variations*: If J is defined by an integral of the form

$$J = \int f(t, y, \dot{y}) \mathrm{d}t,$$

then J has a stationary value if the Euler-Lagrange differential equation

$$\frac{\partial f}{\partial y} - \frac{\mathrm{d}}{\mathrm{d}t}\left(\frac{\partial f}{\partial \dot{y}}\right) = 0 \qquad (13.18)$$

is satisfied, and the Euler-Lagrange of (13.17) becomes:

$$\frac{\mathrm{d}}{\mathrm{d}t}\frac{\partial}{\partial \dot{x}^\lambda}\sqrt{g_{\mu\nu}\dot{x}^\mu\dot{x}^\nu} = \frac{\partial}{\partial x^\lambda}\sqrt{g_{\mu\nu}\dot{x}^\mu\dot{x}^\nu},$$

$$\frac{\mathrm{d}}{\mathrm{d}t}\left[\frac{\frac{1}{2}g_{\mu\nu}\frac{\partial x^\mu}{\partial x^\lambda}\dot{x}^\nu}{\sqrt{g_{\alpha\beta}\dot{x}^\alpha\dot{x}^\beta}} + \frac{\frac{1}{2}g_{\mu\nu}\frac{\partial x^\nu}{\partial x^\lambda}\dot{x}^\mu}{\sqrt{g_{\alpha\beta}\dot{x}^\alpha\dot{x}^\beta}}\right] - \frac{\frac{1}{2}\frac{\partial g_{\mu\nu}}{\partial x_\lambda}\dot{x}^\mu\dot{x}^\nu}{\sqrt{g_{\alpha\beta}\dot{x}^\alpha\dot{x}^\beta}} = 0,$$

The square root term is the light speed at a certain position, hence not a function of time and can be removed, so

$$\frac{\mathrm{d}}{\mathrm{d}t}\left[\frac{1}{2}g_{\lambda\nu}\dot{x}^\nu + \frac{1}{2}g_{\lambda\mu}\dot{x}^\mu\right] - \frac{1}{2}\frac{\partial g_{\mu\nu}}{\partial x_\lambda}\dot{x}^\mu\dot{x}^\nu = 0,$$

$$g_{\mu\nu}\ddot{x}^\lambda + \frac{1}{2}\left[\frac{\partial g_{\lambda\nu}}{\partial x^\mu}\frac{\mathrm{d}x^\mu}{\mathrm{d}t}\dot{x}^\nu + \frac{\partial g_{\lambda\mu}}{\partial x^\nu}\frac{\mathrm{d}x^\nu}{\mathrm{d}t}\dot{x}^\mu - \frac{\partial g_{\mu\nu}}{\partial x_\lambda}\dot{x}^\mu\dot{x}^\nu\right] = 0,$$

$$g_{\mu\nu}\ddot{x}^\lambda + \frac{1}{2}\left[\frac{\partial g_{\lambda\nu}}{\partial x^\mu} + \frac{\partial g_{\lambda\mu}}{\partial x^\nu} - \frac{\partial g_{\mu\nu}}{\partial x_\lambda}\right]\dot{x}^\mu\dot{x}^\nu = 0,$$

$$g_{\mu\nu}\ddot{x}^\lambda + \frac{1}{2}\left(g_{\lambda\nu,\mu} + g_{\lambda\mu,\nu} - g_{\mu\nu,\lambda}\right)\dot{x}^\mu\dot{x}^\nu = 0.$$

This is the equation for a beam of light in the universal (Phantom's) frame. It can be written more compact by introducing the Christoffel symbols (13.13):

$$g_{\mu\nu}\ddot{x} + \Gamma_{\mu\nu\lambda}\dot{x}^\mu\dot{x}^\nu = 0.$$

We can now transform it back to the corresponding equation in our local frame by multiplying it with the inverse of the metric tensor $g^{\mu\nu}$

$$g^{\mu\nu}\Gamma_{\mu\nu\lambda} - g^{\mu\nu}g_{\mu\nu}\ddot{x}^\lambda = 0,$$

and obtain

$$\ddot{x}^\lambda + \Gamma^\lambda_{\mu\nu}\dot{x}^\mu\dot{x}^\nu = 0,$$

$$\frac{\mathrm{d}^2 x^\lambda}{\mathrm{d}t^2} + \Gamma^\lambda_{\mu\nu}\frac{\mathrm{d}x^\mu}{\mathrm{d}t}\frac{\mathrm{d}x^\nu}{\mathrm{d}t} = 0.$$

Compare this equation with the equation for the geodesic in GR, Equation 13.15. Light will follow the same path under GR and in a spatial continuum with varying light speed.

Bibliography

[1] Yavuz Başar and Dieter Weichert. *Nonlinear Continuum Mechanics of Solids.* Springer, 1999.

[2] C. Barceló and G. Jannes. A real lorentz-fitzgerald contraction. http://digital.csic.es/bitstream/10261/3425/3/0705.4652v2.pdf, n,d. Accessed 14-June-2014.

[3] Carlos Barceló, Stefano Liberati, and Matt Visser. Analogue gravity. http://www.livingreviews.org/lrr-2005-12, 2005. Accessed 14-June-2014.

[4] Richard Fitzpatrick. Advanced classical electromagnetism. A graduate level course of lectures. http://farside.ph.utexas.edu/teaching/jk1/lectures/node19.html. Accessed 1-June-2014.

[5] Richard Fitzpatrick. Advanced classical electromagnetism. A graduate level course of lectures. http://farside.ph.utexas.edu/teaching/em/lectures/node31.html. Accessed 1-June-2014.

[6] S. Flügge, editor. *Mechanics of Solids II*, volume VIa/2 of *Encyclopedia of Physics.* Springer, 1972.

[7] Georgia State university, GSU. Gravitational red shift. harvard tower experiment. http://hyperphysics.phy-astr.gsu.edu/hbase/relativ/gratim.html#c2, n,d. Accessed 15-June-2014.

[8] Georgia State university, GSU. Hafele and Keating experiment. `http://hyperphysics.phy-astr.gsu.edu/hbase/relativ/airtim.html`, n,d. Accessed 15-June-2014.

[9] Egil Hylleraas. *Matematisk og teoretisk fysikk*, volume I-IV. Grøndal & Søns forlag, 1950.

[10] Erwin Kreyszig. *Advanced Engineering Mathematics*. JOHN WILEY and SONS, INC, 8 edition, 1999.

[11] Charles W. Misner, Kip S. Thorne, and John Archibald Wheeler. *GRAVITATION*. W. H. Freeman and Company, 1972.

[12] James R. Rice. Notes on elastodynamics, green's function, and response to transformation strain and crack or fault sources.

[13] Bernhard Riemann. On the hypotheses which lie at the bases of geometry. *Nature, Vol. VIII. Nos. 183, 184, pp. 14-17, 36, 37.]*, 1868. Accessed 16-June-2014.

[14] Sir William Thomson. *Mathematical and Physical Papers*, volume III. C. J. Clay and sons, Cambridge University Press, 1890.

[15] Sir Edmund Whittaker. *A History of the Theories of Aether and Electricity*, volume I and II. Philosophical Library, 1951.

[16] Wikipedia. Michelson–morley experiment — wikipedia, the free encyclopedia. `http://en.wikipedia.org/w/index.php?title=Michelson%E2%80%93Morley_experiment&oldid=611540865`, 2014. [Online; accessed 6-June-2014].

[17] Wikipedia. Positronium — wikipedia, the free encyclopedia. `http://en.wikipedia.org/w/index.php?title=Positronium&oldid=632635206`, 2014. [Online; accessed 8-November-2014].

[18] Wikipedia. Radiation pressure — wikipedia, the free encyclopedia. `http://en.wikipedia.org/w/index.php?title=`

Radiation_pressure&oldid=599265826, 2014. [Online; accessed 1-June-2014].

[19] Wikipedia. Ricci calculus — wikipedia, the free encyclopedia. http://en.wikipedia.org/w/index.php?title=Ricci_calculus&oldid=602109094, 2014. [Online; accessed 16-June-2014].

[20] Wolfram Research. Elementary functions. http://functions.wolfram.com/01.10.09.0001.01. Accessed 1-June-2014.

[21] Edward L. Wright. History of the cmb dipole anisotropy. http://www.astro.ucla.edu/~wright/CMB-dipole-history.html, Last modified 2009. Accessed 4-June-2014.

Index

"Rabbit's clever," said Pooh thoughtfully.
"Yes," said Piglet, "Rabbit's clever."
"And he has Brain."
"Yes," said Piglet, "Rabbit has Brain."
There was a long silence.
"I suppose," said Pooh, "that that's why
he never understands anything."
A.A. Milne, Winnie-the-Pooh